ORNAMENTAL AQUACULTURE
Technology and Trade in India

ORNAMENTAL AQUACULTURE
Technology and Trade in India

— Editors —
S. Felix
T.V. Anna Mercy
Saroj Kumar Swain

2013
Daya Publishing House®
A Division of
Astral International Pvt. Ltd.
New Delhi – 110 002

ISBN 9789351241317

Published by : **Daya Publishing House®**
A Division of
Astral International Pvt. Ltd.
ISO 9001:2008 Certified Company
4760-61/23, Ansari Road, Darya Ganj
New Delhi-110 002
Ph. 011-43549197, 23278134
E-mail: info@astralint.com
Website: www.astralint.com

Laser Typesetting : **Classic Computer Services**
Delhi - 110 035

Printed at : **Salasar Imaging Systems**
Delhi - 110 035

PRINTED IN INDIA

Dedicated to.....

I. Joyce Olive Rachel M.F. Sc.

My Beloved Wife

— *S.Felix*

Foreword

The export potential of ornamental fisheries of our country is much higher than the current level. There are roughly 4000 fish breeders involved in the ornamental fish farming sector in our country and Chennai has around 1200 breeders. It is to be reminded that the world trade in this sector presently has recorded 1.5 billion US$. The countries such as Sri Lanka, Singapore and Thailand are involved in the trade at a much higher level than India. India with its rich aquatic biodiversity, 'fisheries - friendly' tropical conditions, available fisheries expertise and with their long association with the sector of ornamental fish farming is yet to exhibit signs of transformation to improve fish production in this area of aquaculture.

Evolving new and innovative strategies, developing adequate aquaculture infrastructures, identification of bottlenecks, implementing Government's schemes to promote cottage, micro, medium and industrial level farming in this sector are the prime need of the hour.

A strategic cooperation of entrepreneurs of this sector needs to be achieved at the earliest with the involvement among themselves and with the Government agencies so as to promote this sector to the newer heights in the days to come. In this time of prevailing economic recessions experienced all over the world, clear business plans, practical consumer–friendly approaches and sound strategies can win the race for this potential sector.

The deliberations of the National Seminar organized by the Fisheries Research and Extension Centre, TANUVAS, Madhavaram on "Technology and Trade Prospects for Ornamental Aquaculture in India" is well documented and published in the form of a book entitled "Ornamental Aquaculture–Technology and Trade in India" by Dr. S.Felix and his team, Professor for the benefit of farming community, students, scholars, teachers, fish breeders, traders and entrepreneurs. I congratulate Dr.S.Felix for his sincere effort to document the technology and trade related aspects of the emerging aquariculture sector which would go a long way to guide the farming sector on the right path of development.

A.G. Ponniah
Director

Preface

Inland and marine fisheries sector in the country is providing livelihood security to millions of people by creating employment opportunities in remote villages and urban areas. It also contributes in eradicating the malnutrition and to boost up the economy of the country. In recent years ornamental fish breeding and culture is gaining considerable importance in the rural areas due to the involvement of women self help groups. Ornamental fish keeping in the house hold for hobby, status, peace and tranquility is on the increasing trend in our country. Annual domestic demand for ornamental trade is estimated to be around Rs. 50 Crores and its growing by 20 per cent annually. This sector is expected to achieve new heights through proper tie up between the producer and markets. This sector gives a scope for utilization of our vast resources in freshwater and brackishwater, which were not utilized fully due to lack of coordinated approach among the various stakeholders.

This current state of affair is largely due to the lack of technical know – how available on the breeding protocol for many species, live –feed- culturing, mass culturing of fish using innovative and intensive culture systems and health management protocols. Further, the existing protocol for the export and import are also not 'farmer – friendly' in our country. Government needs to invest on vital infrastructures such as 'Broodstock facility' and 'Quarantine facility' to streamline and raise the ornamental fish production level to a newer heights in the country.

Under these circumstances, keeping the above facts in mind, a National Level Seminar was organised on "Technology and Trade Prospects in Ornamental Aquaculture" by the TANUVAS to deliberate upon the technical and other related support this sector needs from the Government and also from the Fisheries Institutes. The NFDB, Hyderabad, the DBT, New Delhi, the TNSCST, Chennai, the MPEDA, Kochi the State Fisheries Department and the NABARD, the vital organizations who are working for promoting this sector were made part of this programme for the benefit of farmers/ entrepreneurs.

The book entitled "*Ornamental Aquaculture: Technology and Trade in India*" is the compilation of addresses (of Invited speakers) delivered at the conference by eminent scientists_and fisheries administrators from all over the country. This book would immensely benefit the readers comprising students, scholars, scientists, entrepreneurs and traders. I am indeed greatful to all of them for their valuable contributions.

My heartfelt thanks are also due to *Dr. T. V. Anna Mercy* and Dr. *Saroj K. Swain*, the Co-editors of the book for their commitment in bring out the book.

I hope this documentation, a small step taken in the direction for the development of this potential sector would pave way for greater investment of technical and financial inputs in the days to come to achieve '*aqua rainbow revolution*' using these live aquatic jewels.

S. Felix

Contents

Foreword *vii*

Preface *ix*

Part I
State and National Scenario and Scope for Ornamental Fisheries Sector

1. Scope for Achieving 'Ornamental Rainbow Revolution' in Tamil Nadu 3

A. Sukumaran

2. Opportunities and Challenges in Developing Ornamental Fisheries as a Sun Rise Sector in India for Providing Employment in the Rural Sector and Earner of Foreign Exchange 8

B. Madhusoodana Kurup

Part II
Native Ornamental Fishes of India: Biodiversity, Conservation and Management

3. Indigenous Ornamental Fish Germplam Inventory of the Western Ghats of India: Status and Prospects 39

T. V. Anna Mercy

4. Captive Breeding and Conservation Strategies for Indian Freshwater Ornamental Fish 75

Saroj K. Swain, N. Rajesh and Ambekar E. Eknath

5. Wild Caught Native Freshwater Ornamental Fishes of India 98

J.D. Jameson (Late)

Part III
Ornamental Fish Production Technologies and Management

6. Advanced Grow-Out Systems (Raceways and Lined Ponds) for Intensive Rearing of Ornamental Fishes 115

S. Felix

7. Aquaponics System: Integration of Aquaculture with Vegetable Hydroponics 127

Ravindra D. Bondre

8. Ornamental Arowana 138

V. Sundararaj and D. Yuvaraj

9. Aquarium Fish Health Management, Biosecurity and Quarantine 149

K.M. Shankar

Part IV
Marine Ornamental Fisheries: Biodiversity, Production and Management

10. Sustainable Marine Ornamental Fish Trade: An Indian Perspective 163

G. Gopakumar

11. Marine Ornamental Fishery Resource and its Management at Vizhinjam Coast, Southern Kerala 177

M.K. Anil, S. Jasmine, B. Santhosh, Rani Mary George, B. Raju, C. Unnikrishnan and H. Jose Kingsly

12. Techniques for Mass Production of Two Species of Clown Fish: Clown Anemone Fish *Amphiprion ocellaris* (Cuvier, 1830) and Spinecheek Anemone Fish *Premnas biaculeatus* (Bloch, 1790) 193

M.K. Anil, B. Santhosh, S. Jasmine, Reenamole, G.R. Unnikrishnan, C. and A. Anukumar

13. Hatchery Production Technology for Clown Fish in Tamil Nadu 204

T.T. Ajith Kumar and T. Balasubramanian

14. Induced Breeding Protocols for Ornamental Brackishwater Fishes 214

M. Kailasam, A.R.T. Arasu, J.K. Sundaray, G. Biswas, Premkumar, R.Subburaj and K. Thiagarajan

Part V

Technology Dissemination and Trade

15. The Role of Fisheries Department in Promoting Aquariculture Sector 229

R. Thillai Govindan

16. Factors that Hamper the Ornamental Fish Production Sector in Tamil Nadu 236

N. Mini Sekharan, S. Vivekanandan, S. Asanarr, Soumya Subhra De

Index 249

Part I

State and National Scenario and Scope for Ornamental Fisheries Sector

Chapter 1

Scope for Achieving 'Ornamental Rainbow Revolution' in Tamil Nadu

A. Sukumaran
Commissioner of Fisheries
Government of Tamil Nadu
Chennai – 600 006

A decade long prosperity in the aquaculture sector, experienced in India due to the shrimp farming sector, has slowly now slipped into a kind of transformation mode. However, it may take some more time for its total revival. But, what India in general, Tamil Nadu in particular, needs through aquaculture is micro or small–scale Aquaculture Plan to sustain the livelihood of weaker section of our state. In this context, 'Aquariculture' or 'Ornamental fish culture' provide the viable option or the right employment opportunity for people to earn for their livelihood.

Aquaculture in general, is considered to be a high investment venture, as it demands large size ponds, huge requisite of water, pumps, aerators, cages, etc. But here is an aquaculture practice that can be undertaken in an area, as small as, 500 sq.ft, typically on one's backyard to earn a permanent income.

Global Scenario

Ornamental fish keeping is becoming popular as an easy and stress relieving hobby. About 7.2 million houses in the USA and 3.2

million in the European Union have an aquarium and the number is increasing day by day through out the world. Ornamental fish farming is also growing to meet this demand. The fact is that USA, Europe and Japan are the largest markets for ornamental fish but more than 65 per cent of the exports come from Asia. It is encouraging news for developing countries that more than 60 per cent of the total world trade goes to their economics. While Singapore and other East Asian countries account for 80 per cent of the global trade the main markets are the US, the UK, Belgium, Italy, Japan, China, Australia and southern Africa. The global trade is valued at 282 million US$ at present.

Scenario in India

Although, India is still in a marginal position, its trade is developing rapidly. An estimate carried out by Marine Products Export Development Authority of India shows that there are one million fish hobbyists in India. The internal trade is estimated to be about Rs. 15 crores and the export trade is in the vicinity of US$ 1.0 million. The annual growth rate of this trade is 14 per cent. About 90 per cent of Indian export go from Kolkata followed by 8 per cent from Mumbai and 2 per cent from Chennai. A wide range of availability of species and favorable climate, cheep labour and easy distribution make India, and Tamil Nadu in particular, suitable for ornamental fish culture. India has a miniscule level of Rs.6-7 crores, recording only 0.4 per cent compared to the global trade. This is despite the country's good tropical climate, varied freshwater sources, a long coastline and varied freshwater ornamental fishes. However, the growing demand for ornamental fishes and the growing awareness for farming would change this scenario in India.

Scope in Tamil Nadu

In Chennai, many farmers grow fish in their backyards and sell the stock to firms, which are engaged in the export business. Ornamental Fish culture and trade in Tamil Nadu especially at Kolathoor village on the outskirts of Chennai (Red hills, Devanpattu, etc.) is famous for ornamental fish culture by small–scale producers. There are about 600 families earning their livelihood through ornamental fish culture in Kolathoor and on an average each household in the village earns over Rs. 5,000 to 10,000 per month through ornamental fish farming. About 45 kms from Kolathur,

Gummidipoondi village is another hub of ornamental fish production, where women SHGs have successfully taken up breeding and raising of ornamental fishes to earn their livelihood. On the commercial front, the ornamental fish trade is a growing business with Chennai and Kolkata turning out to be major production and export centres. The domestic trade is confined to medium and small ornamental fish farmers.

Tamil Nadu Fisheries Development Corporation (TNFDC) a Tamil Nadu State Fisheries department undertaking, joined the field in the year 2000. They have initiated activities to rear popular varieties like goldfish, angelfish, mollies and fighters in its farm near Coimbatore. The ornamental fishes are sold in the local markets. Many Indian species like catfish, dwarf and giant gourami, and barbs are popular abroad and fetch good prices. To popularize ornamental fish production and trade in Tamil Nadu, Kolathoor is an excellent example to be prototyped. It provides a good mixture of both cottage as well as commercial–scale production, which largely cater to the export market. More than 200 species of these freshwater fish are bred in different parts of India and others still have to be imported as fry. Majority of the indigenous varieties that fetch good export market come from the Western Ghats and North Eastern region and it is noticed that more than 300 colourful species have so far been identified from these regions. With its tropical climate, which is ideal for ornamental fish culture, Tamil Nadu can become a key player and front runner in the business of ornamental fish culture.

Diversified Ornamental Fish Varieties of India

Two categories of ornamental fish are being marketed–exotic ornamental fish and native fishes of India, which have ornamental value for coloration or behavior. However, exotic fish dominate the domestic market. Already 288 exotic varieties have been recorded in Indian market. According to availability, demand and climatic condition the ornamental fish farmers of West Bengal are mainly engaged in breeding and rearing of common exotic live bearers and egg layers.

Among the preferred ornamental fish, there are common exotic live bearers like guppy, *Poecilia reticulata;* molly, P*oecilia latipinna;* swordtail, *Xiphophorus helleri;* platy, *Xiphophorus maculates* and egg layers like gold fish, *Carrassius auratus;* Koi, *Cyprinus carpio;* tiger barb, *Puntius tetrazona,* Siamese fighter, *Betta splendens;* serpae tetra,

Hyphessobrycon serape and on-growing of some imported fish like silver shark *Balatocheilus melanopterus;* angel *Pterophyllum scalare,* red-tailed black shark *Epalzeorhynchus bicolor;* red finned shark *Epalzeorhynchus erythurus.*

Sometimes the fry of native ornamental fish are collected and sold after domestication. The important ornamental fish in this category include honey gourami, *Colisa chuna;* rosy barb, *Puntius conchonius;* zebra fish, *Brachydanio rerio;* glass fish, *Chanda nama;* "Y" loach, *Batia lohachata.*

Increased Women's Role in the Sector

In southern India, involvement of women in aquaculture is predominantly (82 per cent) limited to collection of wild seed of shrimp in the backwaters during high tides. A large number of poor women are engaged in traditional aquaculture activities and make an important contribution to the rural economy. Nevertheless, in coastal areas, women's participation is mainly confined to marketing of fish, processing, transport and to some extent, net making and mending.

To ensure that women utilize their full potential in profitable activities like aquaculture, it is necessary to provide capacity building support to rural women, which will eventually lead to their empowerment. One such project is 'backyard ornamental fish breeding and management' which has been found to offer immense scope for improving the livelihood of rural women who could use it as an additional alternative income generating venture.

For a Viable and Successful Ornamental Fish Farming the Following should be Given Priority

- ☆ Scientific approaches in managing the aquaculture practices (culture systems, nutrition, disease, water quality, etc).
- ☆ Evolving water source/area based fish breeding programmes.
- ☆ Concentrating on difficult species to breed in their farm sites.
- ☆ Avoiding over production of same species by all entrepreneurs–to avoid crash of price and demand slump in the market.

- ☆ Willingness among fish breeders to evolve need based new strategies.
- ☆ Trying innovative approaches–new and advanced culture systems, management protocols, etc.
- ☆ 'Industry–institute–lead department' based collaborative ventures could be commenced for the mutual benefit of each other.

Role of Tamil Nadu State Fisheries Department for Promoting the Sector

1. Providing Trainer's Training to update the knowledge of technical staff in breeding and other management aspects of ornamental fish farming.
2. Establishment of 'Breeding Centres' in all Carp Hatchery Centres of the Department and also in other suitable places where good quality water source is available.
3. Implementing the projects effectively by extensive extension works with suitable publications (pamphlets, handouts, posters, etc.).
4. To develop an 'Aqua Estate for Ornamental Research Cum Demonstration' facility which can cater to the need of the farming community of the state.
5. Infrastructure for 'Marine Ornamental Fish Hatcheries' can be created at suitable sites to standardise breeding protocol for commercially viable candidate species.
6. Concept of 'Mobile Aquaritech Lab' to be introduced (for testing of soil and water quality, feed quality, fishes for their health, etc)
7. A 'Hi-tech Public Aquarium', accommodating all freshwater and marine (native, exotic) candidate species to be developed in Chennai.

Also through various schemes like IAMWARM, NADP, NFDB and WGDP the Department is assisting the ornamental fish culturists of Tamil Nadu state.

Chapter 2

Opportunities and Challenges in Developing Ornamental Fisheries as a Sun Rise Sector in India for Providing Employment in the Rural Sector and Earner of Foreign Exchange

B. Madhusoodana Kurup

The Vice Chancellor,

Kerala University of Fisheries and Ocean Studies (KUFOS), Kochi, Kerala

Introduction

The ornamental fish trade is a multi-million dollar industry and the global export earning during 2008 was calculated as US$ 344 million while the import was to the tune of US$ 349 million. The ornamental fish keeping is a hobby which is gradually replacing outdoor leisure activities. The soothing effect of aquarium keeping is of immense importance to relieve pressures of today's urban life. The ornamental fish trade has a significant role in the economy of developed and developing countries both as a foreign exchange earner and as a source of employment. This sector assumes special significance due to its huge potential in providing employment to the people hailing especially from rural sector and as a foreign exchange earner. The low production cost and higher returns within a very short time span, involvement of a wide variety of ornamental organisms, ever growing demand for fishes both in the domestic and international markets and the scope for development of new products and accessories to cater to the dynamic needs of the sector are the major attractions, when compared to any other sector. Ornamental fish trade is very dynamic and vibrant in global markets and in order to make the industry more competitive, concerted R&D activities need to be carried out.

Development of captive breeding protocols for the mass production of seeds of the new indigenous species, value addition of fishes with the help of biotechnological tools, improving disease resistance, development of new packing technology for long distance transportation are some of the frontier areas where major breakthroughs need to be achieved. On the other hand, the ornamental fish industry is facing various challenges both globally and regionally. The immediate issues of concern are overexploitation of the wild stock thus causing endangerment to many fishes species such as the most sought barb of Kerala waters *Puntius denisonii*. Incidence of fish diseases, germplasm piracy, loss of biological diversity due to indiscriminate exploitation and habitat loss, potential for fortuitous entry of genetically modified organisms and exotics into the natural ecosystems and its repercussions are other challenges faced by the industry.

Global Trade and Species Involvement

Global trade in ornamental fish is about US$ 6 billion and the entire industry, including accessories and fish feed, is estimated to be worth more than US$ 20 billion. Tropical freshwater fishes contribute 80-90 per cent of the world market, the rest being supported by tropical marine and brackish water species. In freshwater ornamental fish trade, 90 per cent comes from aquaculture and remaining 10 per cent from wild collections. This sector provides employment opportunities to 1.5 million people, mostly from rural sector. Altogether 1600 species are involved in the world trade, 750 species are from freshwater and the remaining belongs to both brackish and marine. USA, Europe and Japan are the major markets and Singapore, Malaysia, Czech Republic, Spain, Indonesia, Japan,

Israel, Thailand, Philippines and Sri Lanka are the major suppliers. FAO reported an annual 10 per cent increase in ornamental fish trade globally, suggesting an overall increase in demand for ornamental fishes.

Both domestic and international markets of this industry are growing progressively. This growing trend of global export of ornamental fishes is obvious from Figures 2.1 and 2.2. Most of the developing countries are the suppliers and the volume markets are developed countries with more than 52 per cent share in the global export market. During 1999-2008, the export grows from US $ 167.6 million to US $ to US $ 343.9 million.Asian countries shared more than 52 per cent of markets, constituted the major supplier.Top exporting countries-Singapore (1st: 20.5 per cent), Malaysia (2nd), Czech (3rd), Japan, Israel, Sri Lanka, Indonesia, Philippines, China, Taiwan.During 1999-2008 fish worth of US $ 245 million to US $ 394 million were imported, notably by USA, UK, Germany, Japan, Singapore, France, Netherlands, Belgium, China, UAE are some of the top importing countries.

The world market of ornamental fish trade is dominated by 30-35 freshwater species and major groups include guppy, molly, angel fish, neon tetra, gold fish, platy, zebra fish, sword tail, danio and discus. Out of total freshwater fish trade, 60 per cent of share is contributed by Neon tetra and Guppy. Marine ornamentals contribute 20 per cent of the trade value.

Status of Asian Countries

Asia continue to be the largest supplier. More than 28 Asian countries are engaged in production and supply while four countries, Singapore, Malaysia, Japan and Thailand shared more than 40 per cent of the global market. Other producers are Sri Lanka, Philippines, Indonesia, China, Taiwan, India and Vietnam. India has a relatively small share in aquarium fish trade worth about US $ 1.7 million during 2008. With proper streamlined production and marketing strategies Asia continues to keep its larger share in the trade (Dey, 2010) (Figure 2.3). Singapore ranks first with 40.1 per cent of the total exports valued US$ 61.4 million. Other major exporters are in the order Malaysia (13.1 per cent), Japan (10.9 per cent), Thailand (9 per cent), China (6.9 per cent) and Indonesia (5.8 per cent). Though, these countries are historically renowned for their indigenous

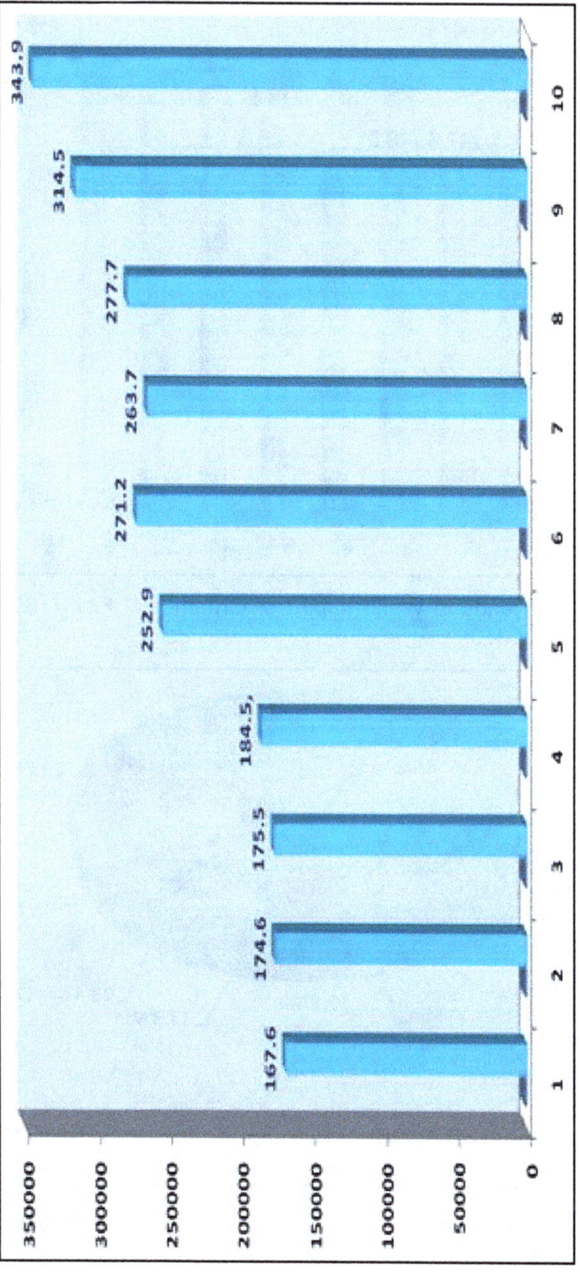

Figure 2.1: Trend of Global Exports of Ornamental Fish (1999-2008) (Value in US$ million) (*Source*: Dey, 2010)

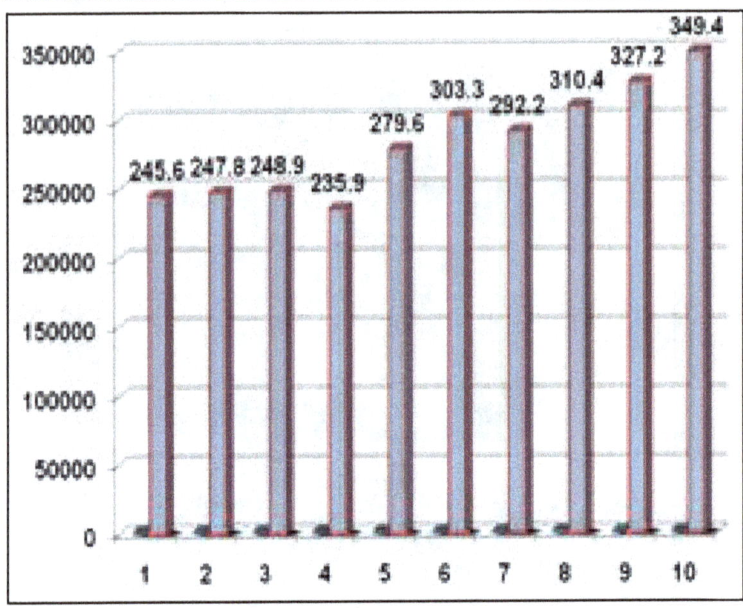

Figure 2.2: Trends in World Import of Ornamental Fish (1999-2008) (*Source*: Dey 2010)

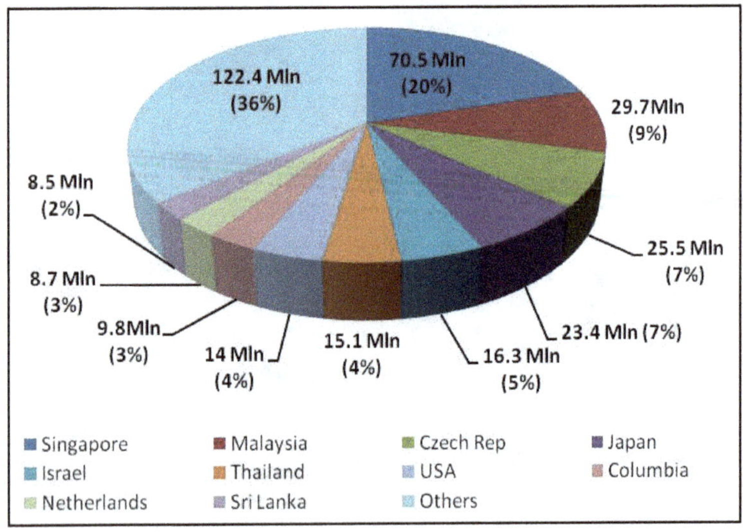

Figure 2.3: Global Exports–Share of Top Ten Countries (2008) (*Source*: Dey, 2010)

germplasm resources and their penetration into global market was achieved with the intensification of production under captive conditions.

Scenario in India

India contributes only around 0.5 per cent of world exports. Indian export during 2000-2008 was 0.8 million to US$ 1.7 million. India exports 100 million fish which includes 124 species, of which 85 per cent of the exports are taking place from northeastern states. India is blessed with a wide array of both marine and freshwater ornamental fish varieties.In 2004, India was ranked 26th in terms of ornamental fish export value in the world compared with its 2nd position it holds in aquaculture. Currently, India contributes only around 0.5 per cent of world exports of ornamental fish and this contribution has remained static. A key reason for this stagnation is that India's exports are currently almost exclusively based on wild-caught freshwater ornamentals and market requirements for these species under current global market is probably saturated. Therefore, to increase India's share of world exports, India must expand its product range and compete aggressively in the market whilst ensuring quality and assured supplies.

Potential available area for aquaculture in India includes 1.9 million hectares of reservoirs, 2 million hectares of tanks and ponds, 1.5 million hectares of bheels and oxbow lakes and 1.4 million hectares of brackish water. Despite India's rich resource of good water and climate essential for breeding and rearing of tropical freshwater ornamental fishes, the appropriate human, material and financial investment to date required to develop the sector is limited.While India's contribution was only 2.5 per cent (US$ 3.8 million) to the total Asian ornamental fish exports (Figure 2.3), India's overall domestic trade in ornamental fish is estimated to be nearly Rs. 15 crores. The values of India's ornamental fish exports during the years 2002-03 to 2006-07 are depicted in Figure 2.4. During the year 2006-2007 our exports valued 5.55 crores against the value of 2.54 crores during 2002-2003. In India, the Western Ghats and North-Eastern States are endowed with plentiful indigenous freshwater fish germplasm which have all the desirable traits for developing as ornamental fishes. Major proportion of the fishes exported from India are freshwater fishes collected from North East and South Western States, followed by exotics and marine fishes. About 400 species of

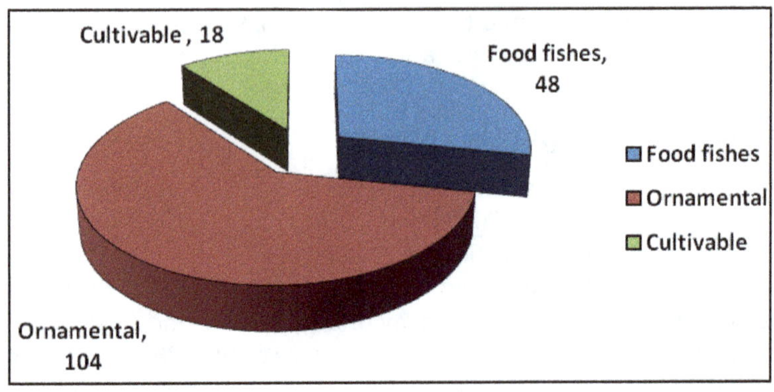

Figure 2.4: Numerical Strength of Ornamental, Food Fishes and Cultivable Freshwater Fishes from Kerala

ornamental fishes are distributed in the Indian seas. Similarly, 376 freshwater ornamental varieties are distributed in the North Eastern region and Southern states. Presently the stock of many of the well sought native freshwater ornamental fishes is under severe threat leading to endangerment due to overfishing and unscientific exploitation for trade purposes. It was also found that only 20 per cent of the catch survived, posing a serious threat to the species. Unlike food fishes, capture based export is not sustainable in ornamental fish trade and it is a matter of concern for the industry.

Potential of India in Overseas Ornamental Fishes Markets

Compared to other Asian countries, the Indian ornamental fish industry is small but vibrant, with potential for tremendous growth and large scale employment generation. India's contribution to the global fish trade is very negligible. Despite having high quality water, unique fish germplasm resources, comparatively cheap labour, availability of qualified man power, the country could not tap the enormous potential available in this sector. The key factors attributing to the success of ornamental fish industry is the development of skilled man power, imparting training on breeding and farming of well sought species in international markets for the bulk production and health management. A concerted attempt is needed in these direction for taping the enormous potential available in the country for ornamental fisheries development.

Rich and Diversified Ornamental Fish Germplasm Resources of India

Marine Ornamental Fishes

Around 10 million ornamental marine fishes are imported annually throughout the world. Around 400 species of marine ornamental fishes are reported in the world which comes under 175 genera in 50 families. India is endowed with vast resource potential of marine ornamentals distributed in the Andaman Nicobar islands, Lakshadweep. Other areas include coastal areas of fringing reefs from Gulf of Kutch to Mumbai, Gulf of Mannar and Palk Bay. More than 50 families consisting of nearly 175 genera and 400 species of ornamental fishes are distributed in the Indian seas. Major families contributing to marine ornamental trade in India includes Pomacentridae, Scorpionidae, Labridae, Apogonidae, Scabridae, Serranidae, Chaetodonidae, Pomacanthidae, Tetrodontidae and Acanthuridae. The species of marine ornamentals coming under different groups are given below.

- ✰ *Damsel fishes: Chromis caeruleus, Dascyllus aruanus, Chromis chrysurus*
- ✰ *Parrot fishes: Scarus psittacus, S.bataviensis*
- ✰ *Surgeon fishes: Acanthurus triostegus, A.lineatus*
- ✰ *Wrasses: Halichoeres hortulanus, Stethojulis albovittata*
- ✰ *Goat fishes: Mulloidichthys samoensis, Parupeneus macronemus*
- ✰ *Butterfly fishes: Chaetodon auriga, C. trifasciatus*

Fish Germplasm Resources of Western Ghats

The Western Ghats, extending for over a length of 1600km, lying parallel to the coast hardly 50km away, form one of the magnificent escarpments of late Tertiary age. Except for a short gap near Palghat, it is unbroken throughout its length. The marked diversity of landscape, the youthful character of the rivers, the precipitous escarpments, the narrow gorges and the relative high elevation of the plateau compared to the plains are the key geographical features of these Ghats. World conservation monitoring centre has identified Western Ghats as one of the important freshwater biodiversity hotspots. Western Ghats forms a major source for varieties of freshwater fishes in India. There are around 170 fish species comes under 64 genera and 32 families. Out of this 104 species are potentially usable as ornamentals, 48 species are food fishes and 18 species are cultivable (Figure 2.4).

Among the 104 ornamental species, 10 species such as *Puntius denisonii* (Red line torpedo fish), *Puntius arulius* (Aruli barb), *Puntius conchonius* (Rosy barb), *Puntius filamentosus* (Filament barb), *Puntius ticto*(Ticto barb), *Puntius vittatus* (Koolie barb), *Puntius fasciatus* (Melon barb), *Parambassis thomassi* (Glass fish), *Horabagrus brachysoma* and *Horabagru nigricollaris* (Yellow cat fishes) have already secured a position in the National and International markets as ornamental fishes while the rest of the species have tremendous potential for their introduction as ornamental species. Major ornamental fish species recorded from the Western Ghats are displayed in Plate 2.1. Captive breeding and seed production technology of most of these fishes are not yet standardised and this forms the major bottleneck for their introduction as ornamental species. Of the 170 species recorded from the Western Ghats, 4 species are exotic and alien in origin

The freshwater fish diversity of Kerala is facing serious threats. Biodiversity threats in the form of diverse types of human interventions are the main reason for the alarming decline of fish populations in most of the rivers. The stock of some of the freshwater ornamental fishes which are rare in the international markets are also suffers due to stock depletion due to their unsustainable exploitation to cater the requirement of the export.

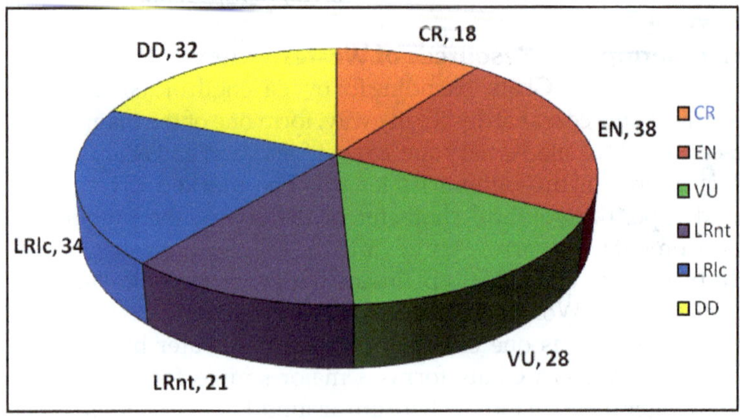

Figure 2.5: Biodiversity Status of Freshwater Fishes from Kerala

Plate 2.1

Puntius denisonii

Osteochilus nashi

Puntius arulius

Labeo nigrescens

Neolissochilus wyanadensis

Puntius ophicephalus

Puntius filamentosus

Chela dadiburjuor

Puntius jerdonii

Barilius bakeriz

Puntius ticto

Brachydanio rerio

Danio malabaricus

Mesonemacheilus menoni

Esomus thermoicos

Mesonemachilus guenthe

Schistura pambarensis

Lepidopygopsis typus

Oreonectes keralensis

Glyptothorax anandalie

Microphis cuncalus

Sicyopterus griseus

Macrognathus guentheri

Mastacembeles armatuz

Some of the gold fish Varieties

Calico Oranda

Brown Oranda with Red Head

White Oranda with Red Cap

Red pearl dragon-eye

Grape eye

Blue dragon-eye

Varieties of Koi Carp

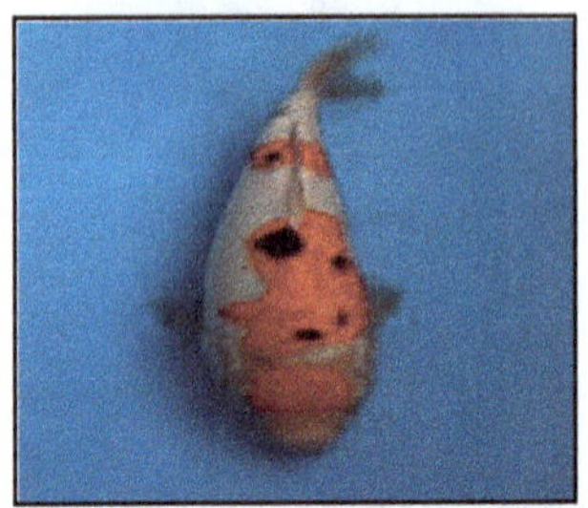

Different varieties of Guppy

Ornamental Fishes of North East India

The northeast region of India, comprised of the states of Arunachal Pradesh, Assam, Manipur, Meghalaya, Mizoram, Nagaland, Sikkim, and Tripura, is blessed with rich biodiversity and fisheries resources. About 80 per cent of ornamental fishes from India to International market are exported via Kolkata Airport, of which the lion's share (more than 80 per cent) is contributed from North Eastern Region. In India the North Eastern states are taking a main role in the ornamental fish market; contributing around 85 per cent of the total market and the rest comes from the southern states of India. The North Eastern region homes around 300 native ornamental fish out of the 806 freshwater fishes found in India.Already 217 fish species belonging to 136 genera has been identified in Assam, of which about 150 species have been reported to be of ornamental value and for more than 50 species, overseas demand has been established. Some of the commercially important species are:

Chitala chitala, Notopterus notopterus, Gonialosa manmina, Gudusia chapra, chela cachius, C. laubuca, Salmostoma bacaila, Brachydanio rerio, Danio aequipinnatus, D. devario, D. dangila, D. regina, Esomus danricus, Aspidoparia morar, Amblypharyngodon mola, Puntius chola, P. conchonius, P. gelius, P. phutunio, P. sophore, P. terio, P. ticto, Osteobrama cotio cotio, Rasbora rasbora, Barilius barila, B. bendelisis, Acanthocobitis botia, Botia histrionica, B. berdmorei, B. derio, Lepidocephalus guntea, Mystus vittatus, Mystus cavasius, Rita rita, Gagata cenia, Hara hara, Ailia coila, A. punatata, Pseudotropius atherinoides, Clarias batrachus, Heteropneustes fossilis, Chaca chaca, Xenontodon cancila, Aplocheilus panchax, Monopterus cuchia, Chanda nama, Pseudambassis ranga, P. lalia, P. baculis, Badis badis, Nandus nandus, Glossogobius giuris, Anabas testudineus, Colisa fasciatus, C. lalia, C. sota, Ctenops nobilis, C. stewarti, Channa orientalis, C. punctatus, C. barca, Macrognathus aral, M. pancalus, Mastacembelus armatus, Tetraodon cutcutia etc. (Directorate of Fisheries, Assam). While the major species are *Channa barca, Channa aurantimaculata, Puntius gelius, P. manipurensis, P. shalynius, Botia spp., Sisor rhabdophorus* and *Erethistes spp.*

Ornamental Fish Industry of Asian Countries

Singapore ranks first with 40.1 per cent of the total exports valued US$ 61.4 million. Other major exporters are in the order Malaysia (13.1 per cent), Japan (10.9 per cent), Thailand (9 per cent), China (6.9 per cent) and Indonesia (5.8 per cent) and they are the major competitors of India in the export market. The status of these countries as exporters and suppliers is dynamic with notable changes in their competing position.

Singapore

The ornamental fish industry of Singapore comprised of 400 species and 1000 varieties producing from 70 farms (160ha.). Almost 300 million fishes are produced per year. Singapore is known as Super market for ornamental fisheries.During 2007-2008 the expoert contribution declined from 22.9 per cent to 20.5 per cent.Singapore imports fish worth of US $ 24.5 while her export was 43.5 US $. The export is depended on fish imported from other countries such as Malaysia. In 2004 India supplied 2.8 per cent of the total exports to Singapore endowed with a very important trading hub and excellent logistic and air cargo service. Among various strengths, excellent infrastructure, marketing net work and Govt. support are worth mentioning.

During the period 1995–2004, Singapore's export plummeted from 40 to 20 per cent. In 2004, Singapore exported around 280 million ornamental fish worth US$51.6 million. In the same year Singapore accounted for 21 per cent of world market in terms of value (FAO, 2006). They depend on fishes imported from other countries mainly from Malaysia. India has supplied 2.8 per cent of the total exports from Singapore in 2004. They imported around US$ 380 000 worth of ornamentals from India in 2004 for value addition and re-exports. Despite this downturn, Singapore, however, is still likely to be a significant player as a supplier of ornamental fish in the near future. It is a trading hub with excellent logistic and air cargo services. The government has a comprehensive R&D and market development programme under the Agriculture Veterinary Authority (AVA) and exporters have relatively easy access to other species in the region to provide a comprehensive supply list.

Malaysia

Malaysia has doubled its world market share in a decade from 4 per cent in 1994 to 8 per cent in 2004, and was aiming to capture 13 per cent. During this period, their world ranking improved from 10th place to 4th. Supply data suggest that 70-80 per cent of Malaysian production is exported to Singapore. New airport facilities in

Malaysia and the promotional activities for the sector by the Malaysian government have resulted in an increase in direct exports from Malaysia. The production trend has shifted from wild caught to cultured in recent years (Figure 2.6).

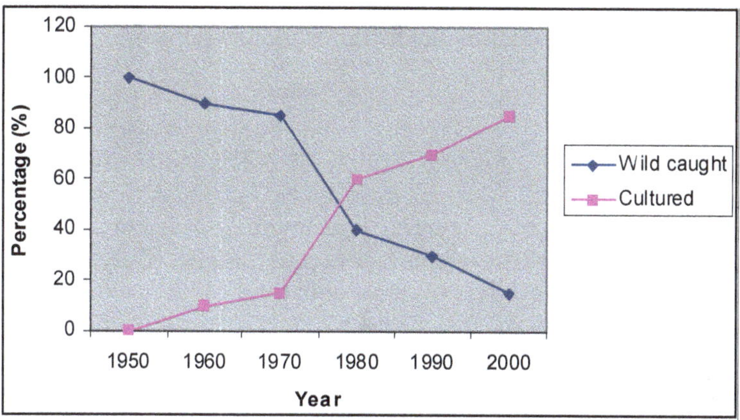

Figure 2.6: Production Ratio of Wild Caught and Cultured

Thailand

Thailand has also rapidly penetrated the main Western markets and is regarded by Malaysia and Singapore as a major competitor. Based on FAO export data, Thailand's exports increased from a mere US$ 550 000 in 1994 to around US$ 10 million in 2004 achieving a staggering growth of 75 per cent/year. Government of Thailand is offering a lot of incentives such as land, infrastructure and finance to promote the export.

Sri Lanka

Sri Lanka was historically renowned for its marine ornamental fish and invertebrates. In recent years, they have, through state intervention, successfully embarked on a programme to increase their share of freshwater ornamental fish markets through government intervention. Presently Sri Lanka is also renowned for its freshwater fish, especially guppies. The countries success in market penetration is evident in data from the last reporting decade as exports increased from US$ 0.95 million in 1994 to nearly US$ 7.4 million by 2004, representing an average annual increase of 68 per cent/year. Building on their renowned global position for marine ornamentals facilitated this market penetration. Sri Lanka's global market share increased from 2 per cent in 1994 to a peak of 5 per cent in 1999 before declining to 3.2 per cent in 2004. During this period, their world ranking moved from 21st to 10th. They had a major switch over from wild caught fishery to farmed stock. They are renowned for the freshwater ornamental fishes, especially guppies.

China

During 1977-1986 the export from China increased from 0.5 million US $1987-2006 10.5 million US $.The value of exports accounted for only 3.7 per cent to the global figures, with an annual growing at a rate of 10 per cent. China imported fish worth of 8-12 million US $ (2000-2008).China lack unique varieties of ornamental fishes but is well developed as far as aquarium gadget technology is concerned. China can be a prospective market for the indigenous ornamental fishes of India. The possibilities of joint ventures between India and China for production of sophisticated gadgets for the ornamental fish production systems in India needs to be explored.

Captive Breeding

At present cultured marine ornamental fishes contribute only 1-2 per cent of the trade worldwide. There are around 84 marine species reared under captivity but the species that are reliable to breed are very few. There are twelve major groups of marine ornamental fishes including common clown (*Amphiprion percula*), false clown (*A. ocellaris*), orange anemone fish (*A. sandaracinos*), three spot *damsel* (*Dascyllus trimaculatus*), humbug damsel (*D. aruanus*), blue damsel (*Pomacentrus caeruleus*) *and* peacock damselfish (*P. pavo*) were successfully bred under captivity. The feed management is the crucial factor in marine ornamental fish keeping and breeding which plays a key role in maintaining the water quality. Successful marine ornamental fish breeding is now possible mainly due to the recent scientific advancement made on various aspects like biological filtration and also the advent of an array of aquarium gadgets.

In the northeast, researchers of various organizations like Central Inland Fisheries Research Institute (CIFRI), Guwahati University, Dibrugarh University, Assam University, College of Fisheries (Assam Agricultural University) etc. have done some pioneering works. The North Eastern Regional Centre of CIFRI has been conducting research on various aspects on indigenous ornamental fishes of this region including cataloguing of potential species, methods of collection from the wild, conditioning of wild fishes and laboratory rearing, etc. Fish species successfully bred and reared at Central Institute of Freshwater Aquaculture (CIFA), Bhubaneswar are *Brachydanio rerio* (Zebra fish), *Brachydanio frankei* (Leopard danio), *Puntius conchonius* (Rosy Barb), *Colisa fasciata* (Banded gourami), *Colisa lalia* (Dwarf Gourami), *Parlucioma daniconius* (Black line rasbora), *Esomus barbatus* (Flying barb), *Danio aequipinnatus* (Giant danio), *Danio devario* (Torquoise danio), *Puntius sophore* (Sophore barb), *Puntius ticto* (Two spot barb) and *Badis badis* (Chameleon fish), *Puntius filamentosus* (Filament barb). Some of the species like *P. denisonoi* and *P. tamraparnei* are in pipeline.

Domestic and Export Market Trade

Although no definitive surveys on the domestic demand have been conducted for ornamental fish, available information suggests that the domestic market may be huge. According to Lukram (2005) the domestic ornamental fish market is worth around Rs. 500 million and the demand is increasing at 20 per cent annually. Other reports suggest annual growth as high as 40 per cent/yr (International Market News 27 Sep 2005), while another states the sector is worth Rs. 10 crores.

This optimism can be further confirmed by encouraging shifts in the number of households and concomitant increasing wealth. Since the early 1990's the total number of households has increased by 32 per cent with new household formation being a potential source of demand for ornamental fish. The domestic sector in India, therefore, potentially offers a huge untapped market in the coming decades.

Challenges

Ornamental fish trade in the international market is facing numerous problems. There is no proper infrastructure and technical expertise for mass production of prime species in demand in the global market. Farmers are not well aware about the potential of ornamental fish trade in the domestic and international markets. At present there is no assured supply of quality fishes in bulk quantities as per requirements. Industry fully depends on wild caught fishes. There is no mechanism to implement quality assurance programme for export of ornamental fishes.

Like food fish trade there is no organisation in ornamental fish sector, to ensure health quality standards and animal welfare conditions which is stipulated by importing countries especially Europe and USA. Stock depletion started occurring for native species which have great demand in overseas markets. The country has no mechanism to ensure quality brood stocks of prime species of world market. There is no diversification in the product mix and introduction of new varieties. There is no institute to exclusively serve as platform to meet the requirements of emerging ornamental sector.

Most important challenges requiring immediate attention are :

1. Infrastructure and technical expertise for mass production of prime species.
2. Demand in the global market
3. Assured supply of quality fishes in bulk quantities
4. Industries fully depending on wild caught fishes
5. Mechanism to implement quality assurance programme for export
6. Institutes to ensure health quality standards, animal welfare conditions, stipulated by importing countries especially Europe and USA
7. Market requirement of native ornamental fishes in global market becoming saturated
8. Mechanism to ensure quality brood stocks of prime species for world market
9. Introduction of new varieties and diversification in the product mix needs to be curbed
10. Institutes to exclusively serve as platform to meet the requirements of emerging ornamental freight and transport
11. Trade negotiation with airlines-to make the rates competitive
12. High cargo rates compared to other Asian countries
13. Lack of logistics and non-availability of direct flight connection to various European and USA markets
14. Negotiations on export barriers and stringent export regulations
15. Tough export procedure

16. Lack of market information
17. Difficulty in consignment filling
18. Other export barriers like possible introduction of undesirable species, diseases, etc.
19. Regulations of import becoming more stringent
20. EU markets being increasingly regulated in respect of the following :
    - ☆ Fish health
    - ☆ Animal welfare
    - ☆ Transportation
    - ☆ Phyto- sanitation
    - ☆ Green certification
    - ☆ Sustainable resource management

Fish health is the key regulatory factor governing trade in the context of the following regulations being prepared and imposed

- ☆ 'Aquaculture health Directive of EU' -Animal health status and certification requirement for importation of tropical and coldwater fish
- ☆ 'EU directive'–Implementation of OIE (official international des Epizootics) guideline
- ☆ 'EU regulation' - New transport regulation

  Factors that are assessed

  1. Protection of live animals during transport
  2. Handling
  3. Welfare
  4. Health status of animal
- ☆ Imports of genetically modified fishes are banned on EU& Singapore. However, they are permitted in USA and China.
- ☆ Green &Eco certification is particularly insisted in USA.

Initiatives by the Government of Kerala: A Model System for Adoption

In due recognition of the socio economic importance of this sector, Govt.of Kerala have taken several initiatives for triggering the ornamental fish export striving to achieve a significant share in the world trade of ornamental fishes, including building up of infrastructure by commissioning of the first Aquatechnology park in the country by the Kerala Aquaventures International Ltd.(KAVIL)- a public private partnership company with its export hub at Cochin and satellite farms across the country. This company, for the first time in the country imported the brood stock of 55 species of ornamental fishes and the F1 and F2 generation of the stock is produced and supplied to the farmers for multiplication targeting the export of bulk volumes to international destinations. Bulk volumes of fishes have already been supplied to various international destinations. The company will hopefully generate tremendous employment opportunities in various facets of ornamental fish industry such as breeding and mass production of fishes in satellite and homestead farms, aquarium plant culture, trading of fish, manufacture and marketing of aquarium and accessories, fish feed, etc., medicines, etc. The first public private Ltd. Company has been formed by the joint participation of Govt.of Kerala and other private stake holders. The Aqua Hub site contains a dedicated company facility having provisions for quarantine facilities and a R&D complex. The Company provides common services such as water, electricity, laboratory services, and quarantine, R&D and extension services to independently owned and managed, on site, ready-to-use fish conditioning and packing units. These sales and marketing units are linked to downstream satellite farmers and their associated homestead producers across the 14 districts of Kerala. The ATP also provides opportunities for a limited number of wild collectors of freshwater fishes and hence supports alternative livelihood opportunities. The ATP comprises satellite farms built on government and private lands. These farms have an association with homestead farms who will outgrow fish for satellite farms on a buyback scheme. The fish reared on satellite farms brought to the Aqua Hub in climate-controlled vehicles where they are reconditioned and packed for sale to national and international markets.

The ATP is facilitated with a CPU, encompassing two main sections; an external facility containing conditioning concrete tanks for conditioning incoming fish for 7-10 days and an internal packing house containing glass tanks for final quality control checks and a packing area for boxing the fish conditioned for sale. The packing facility has a climate control room to maintain the fish at around 20-22°C. Each CPU is enclosed within a compound wall to enhance biosecurity and also providing added security. Government of Kerala has embarked on a structured programme to develop the ornamental fish sector through farming. The Aquatechnology will have very strong interventions in taping the full potential available in the sector by importing the brood stock of prime ornamental fish species which most sought from time to time and also establishing marketing linkages with international trade destination points for the supply of bulk volumes with out compromising the quality of the fishes exported.

Strategies for the Better Future of Ornamental Fish Trade

1. Mass production and supply of high quality fishes for meeting the demand of local and export markets.
2. Standardize the captive breeding technology of indigenous marine and freshwater fishes
3. Make use of the rich resources of water, ideal climatic conditions, educated human capital for the mass production of ornamental fishes
4. Develop an international trading hub and excellent logistic and air cargo services
5. Setting up of 'one–stop–shop' under public private participation to serve as a platform for meeting all the requirements of ornamental industry such as production, marketing, quarantine, certification, etc.
6. Bring the farmers and entrepreneures under the purview of the new institution to generate maximum income and employment
7. Implementation of CITES certification
8. Zoning of potential ornamental fish culture area, develop them in to 'Ornamental Fish Estates' and provide all necessary facilities for mass production
9. Develop cluster farming system by making linkages with the main production farm and homestead, village, public, private ponds available in a region.
10. Homestead, village, public, private ponds etc. of a locality can serve as the fattening centres of the satellite farms and buy back arrangement shall be done

11. Farm accreditation, code of practices, legislation, etc. can be done to maintain quality standards targeting the international export
12. Strengthening of the institutional support by intensification of research and development, training programmes, extension activities, etc.
13. Capacity building of fish collectors in non - destructive collection techniques
14. Intensification of research gaps in:
    (*a*) high technology production system and post harvest handling
    (*b*) enhancing production efficiency and product quality
    (*c*) biotechnological application to develop new strains, improving shape and vibrancy, morphology, disease resistance, growth enhancement and maturation
    (*d*) development of better and competitive logistic support and marketing net work

Future Planning

- ☆ By 2016 the export requirements would be 2.5 billion fish worth of US $ 450 million
- ☆ Increase in demand for fish would be 5 per cent per annum
- ☆ Majority of aquarium are freshwater, scope of increasing production in Asia is immense
- ☆ Farm accreditation, code of practices, legislation, etc can be done to maintain quality standards targeting the international export
- ☆ Develop better marketing networks, extend support from Govt. for the mass production of ornamental fishes sought through aquaculture
- ☆ The possibilities of joint ventures between India and China for production of sophisticated gadgets for the ornamental fish production systems in India may be explored.
- ☆ Efforts shall be made to develop the ornamental fish industry of the country in tune with the neighbouring countries such as Malaysia, Sri Lanka, Indonesia and Singapore, etc.

- Adequate infrastructure such as production and export centres, transport facilities, subsidized air freights, development of backyard units as fattening centres by providing financial assistance, supply of high quality seeds for farming and fattening are some of the important infrastructural requirements for developing the ornamental fish industry in tune with the other Asian countries.

Conclusion

Development of ornamental fish industry would benefit the human to earn a living. Breeding of native and exotic fishes for the domestic market fetches an appreciable price. As the breeding units do not require much investment and special infrastructure facilities, it can be easily adopted by poor farmers and unemployed youths. If this fishery sector is nurtured as a potential industry, it would generate employment and flourish the economy of India through export to earn huge foreign exchange. With the mastery of controlled captive breeding and initiatives to create a large network of breeding centres, India has the potential to become a major player in global ornamental fish trade in the coming years. In order to sustain the growth it is necessary to shift the focus from capture to culture based development. Organised trade in ornamental fish depends on assured and adequate supply of demand, which is possible only by mass breeding.

- Achieve 5 per cent of the world export market during the coming years and gradually increase the achievable target of 10 per cent by 2015
- Develop an international trading hub at Cochin with excellent logistic and air cargo services
- Penetrate in to world market by carving out with a blend of farmed stock of both exotic and indigenous fish germplasm resources
- Mass production and supply of high quality fishes for meeting the demand of local and export market
- Concerted research and development in captive breeding technology of indigenous marine and freshwater fishes.

Part II

Native Ornamental Fishes of India: Biodiversity, Conservation and Management

Chapter 3

Indigenous Ornamental Fish Germplam Inventory of the Western Ghats of India: Status and Prospects

T.V. Anna Mercy

Department of Fishery Biology, College of Fisheries, Kerala University of Fisheries and Ocean Studies, Panangad, Kochi – 682 506

The Global Scenario

Ornamental fish keeping is one of the most popular hobbies in the world today. The growing interest in aquarium fishes has resulted in steady increase in trade globally. The trade has a turnover of US $ 5 Billion and an annual growth rate of 8 percent thus offers a lot of scope for development. The country at the top of the export list is Singapore followed by Hong Kong, Malaysia, Thailand, Philippines, Sri Lanka, Taiwan, Indonesia and India. The largest importer of ornamental fish is the USA followed by Europe and Japan. The emerging markets are China and South Africa. Over US $ 500 million worth of ornamental fish are imported into the USA each year.

India's share in ornamental fish trade is estimated to be Rs. 5.5 crores (2006-07) which is against the value of 2.54 crores during 2002-03. The domestic market is buoyant too, which is mainly based on domestically bred exotic species. The overall domestic trade in this sector is over 15 crores and is growing at a rapid pace of 20 per cent annually. The earning potential of this sector has hardly been understood and the same is not being exploited in a technology driven manner. Considering the relatively simple techniques involved, this sector has the potential to create substantial job opportunities, besides helping export earnings for the country.

The Importance

On a broad basis the level at which they operate we can classify the benefits, essentially into four broad categories.

1. As a hobby
2. As a trade
3. As an art
4. As a scientific entity
5. As a tool for socio economic development

One must understand that the delimiters that are used to divide the above four divisions are not explicit and there can be overlaps. Here the importance of trade is discussed in detail.

As a Trade

Trade occurs at two different levels. They are domestic and export markets.

There are many trade offs to be weighed before one decides the playing field. The demand is huge enough to accommodate more players. As in all cases it has its disadvantages but on the outset the advantages outnumber the disadvantages by a huge margin.

The Domestic Market

The domestic market of ornamental fishes is mainly dominated by exotic fishes. Majoring (95 per cent) of the fishes are the predecessors of exotic species which were brought to India many years ago. These fishes are tank bred in small scale or back yard hatcheries or breeding units. But due to the lack of adequate infrastructure and quality brood stock, they are not in a position to produce new varieties which have demand in the International market.

The Export Market

Major portion of the ornamental fishes exported from India are freshwater fishes collected from the wild. India is blessed with a rich diversity of freshwater fishes both in the Western Ghats and North East Hills, which are hotspots. The Western Ghats that extends from Kanyakumari to River Tapti for an extend of 1600Km. Numerous rivers originating from this mountains store a rich diversity of endemic fish fauna.

Ornamental Fish Resources of the Western Ghats

A detailed study on the ornamental germplasm inventory of the Western Ghats of India was carried out by College of Fisheries, Panangad, Kochi during the period from 2000-2004 as part of the National Agricultural Technology Project (NATP) entitled "Germplasm inventory, evaluation and gene banking of freshwater fishes of India", which is the first of its kind in India. Under the project a database of the indigenous ornamental fishes of the Western Ghats of India was prepared. Out of the 300 and odd species of freshwater fishes of the Western Ghats, 200 species are endemic. 155 species are found to have the desirable qualities of ornamental fishes of which 117 are endemic to the Western Ghats.

Hitherto Untapped Resources?

Even though India has a rich resource of freshwater ornamentals, it was not properly exploited due to the following reasons.

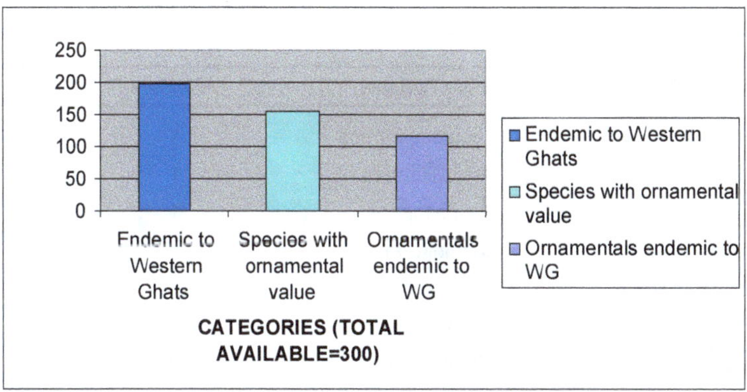

Figure 3.1: Ornamental Germplasm Inventory of the Western Ghats

1. Unawareness of the rich aquatic resources.
2. Proper database of the resources was not available.
3. Ornamental fish export from India is solely depending on the wild caught species, that too is restricted to a few species only.
4. Captive breeding technology of the indigenous ornamental fishes are not developed so that the demand for lump sum quantities could not be satisfied by the suppliers.
5. Due to lack of scientific knowledge in the collection, handling and packing the fish collectors are unable to provide quality fishes.
6. The global ornamental fish traders normally place orders for large quantities. They look for uniform sized disease–free- fishes. It is practically impossible to cater to this demand from wild catches.

The Present Scenario

Many of the problems detailed above have been resolved now to some extent. But we have miles to go to have an organized sector of ornamental fish export. Presently a database on the germplasm of ornamentals of the Western Ghats is available. A book entitled "Ornamental fishes of the Western Ghats of India" has been published by the National Bureau of Fish Genetic Resources, Lucknow. Captive breeding technology for a dozen of indigenous ornamental fishes have also been developed under the NATP project of Indian Council of Agriculture Research.

Sl.No.	*Fish Groups*	*Numbers*
1.	Osteoglossiformes	1
2.	Anguilliformes	1
3.	Clupeiformes	1
4.	Cypriniformes	38
5.	Siluriformes	10
6.	Cyprinodontiformes	1
7.	Perciformes	16
8.	Tetraodontiformes	1

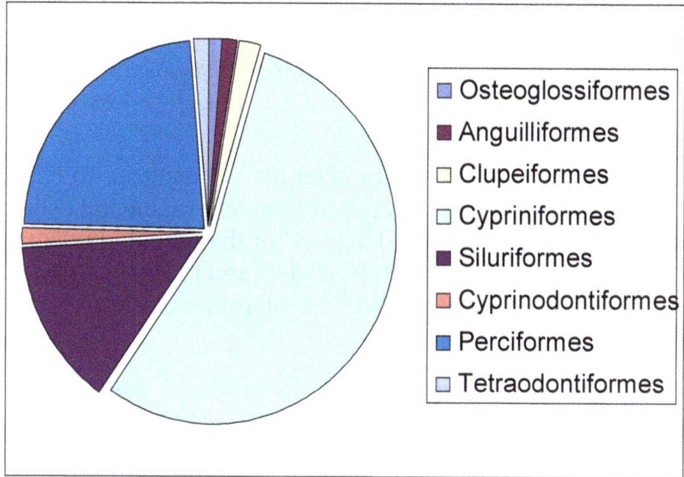

Figure 3.2: Taxonomic Status of Ornamentals of the Western Ghats

Significance of *Puntius denisonii*, 'Miss Kerala'

But the most popular fish in the international market is a barb that is endemic to the Western Ghats which is popularly known as "Miss Kerala". The species contributes to 40-50 per cent of India's ornamental fish export. This fish is so much in demand that many scientists and farmers from different parts of the world have been trying to develop captive breeding technology for this species. Under a project of Marine Products Export Development Authority, College of Fisheries could develop captive breeding technology for this species, which is a breakthrough in the field of native ornamental fish breeding. The technology is standardized so that this fish can be produced in large quantities for export. The technology may be now dissiminated to others for production.

As mentioned above, all the ornamental fishes exported from India are caught from the wild. Due to the indiscriminate exploitation from the wild, many of the species has become endangered. A stock assessment study made on the much sought after fish *Puntius denisonii* has revealed that this fish has become endangered due to the lack of scientific knowledge while collecting, handling and packing the fish. Based on the recommendations of this study and to address some of the issues a workshop on sustainable harvest and green certification of wild caught indigenous ornamental fishes was conducted by Marine Products Export Development authority (MPEDA) in association with United Nations Conference on Trade and Development (UNCTAD) in October 2008. Accordingly, the guidelines have been formulated and it was officially released during February 2010.

The extensive work undertaken by our institute to collect, identify and document a varieties freshwater ornamental fishes of Western Ghats is listed in the Chapter for the benefit of scientists, scholars, fish breeders, exporters, traders and policy makers. This comprehensive document will be of immense help for the stakeholders of the sector.

Barilius Bakeri (Day)

Common Name

Malabar baril (English)

Fin Count

Dorsal ii-iii 10; Anal ii-iii 14; Pectoral i 14; Ventral i 8.

Distinguishing Characters

Body deep, its depth 2.9 to 3.2 times in standard length. Mouth moderate; jaws short, maxilla extends to below middle of orbit; barbels absent. Dorsal fin inserted in advance of anal fin, extending to above fourth anal fin ray. Scales moderate, with few radii; lateral line with 37 or 38 scales; predorsal scales 16. Tubercles large and well-developed on snout and lower jaw.

Colour and Size

Grayish becoming white on abdomen; a row of large bluish spots along the flanks. Dorsal, anal and pectoral fins with dark gray bases, their edges white. It attains a maximum size of 15cm.

Food and Feeding

An omnivorous fish, readily accepts anything its mouth can hold, not at all fussy about food. Prefers insect larvae. Feed from the surface of water column.

Sexual Dimorphism and Breeding

Males develop intensive colouration during breeding time. Colour of fin edges intensifies during maturation. Females develop bulged belly. Captive breeding has not been attempted so far.

Aquarium Requirements

It is a hardy fish. Moves very fast in the tanks hence requires larger tanks.

Behaviour in Captivity

Compatible, lovely and hardy species.

Chela fasciata Silas

Common Name

Malabar hatchet chela (English)

Fin Count

Dorsal ii 7; Anal iii 14-15; Pectoral i 8-9; Ventral i 5-6

Distinguishing Characters

Body greatly compressed, its depth 3.8 to 4.3 times in standard length. Head slightly turned upwards. Mouth small, obliquely directed upwards, its cleft not extending to below front edge of eye.

Pectoral fins long, extend much beyond origin of anal fin; outer ray of pelvic fin greatly elongated, extends beyond origin of anal fin. Lateral line complete, with 33 or 34 scales; lateral transverse scale rows 6/1/1-1½; predorsal scales 18.

Colour and Size

Upper half of body grayish, the scales with dark edges; lower half and belly lighter in colour; a dark broad lateral stripe on sides, commencing just behind eye and runs along middle of body to about base of caudal fin; a well-defined black supraanal streak present, so also a subpeduncular stripe; a mid-dorsal stripe from occiput to dorsal fin. Fins grayish white. It attains a maximum size of 5cm.

Food and Feeding

It is an omnivorous fish, readily accepts anything its mouth can hold, not at all fussy about food. Prefers insect larvae. It never takes food from bottom. Hence floating feed is preferable.

Sexual Dimorphism and Breeding

Males become more colourful and fins become darker during breeding season. The eggs are adhesive and remain attached to the roots of floating plants. Plants are essential in a breeding tank.

Aquarium Requirements

Less demanding fish, but prefers clear and aerated water.

Behaviour in Captivity

It is a popular aquarium fish, moves faster along the mid water column. A good candidate species for planted aquarium.

Danio malabaricus (Jerdon)

Common Name

Malabar danio (English)

Fin Count

Dorsal ii 10-13; Anal iii 12-16; Pectoral i 14; Ventral i 7

Distinguishing Characters

Body elongate and strongly compressed, its depth 3 to 3.5 times in standard length. Head length 3.8 to 4.3 times in standard length; snout length 3.5 to 4, eye-diameter 3.3 to 3.8, both in head length. A weak preorbital spine, directed backwards, from lachrymal bone.

Mouth small, directed upwards; barbels two pairs; rostral barbels rudimentary, the maxillary pair usually vestigial. Dorsal fin inserted well in advance of origin of anal fin, its posterior half extending to over anterior anal fin rays. Caudal fin forked. Scales moderate-size; lateral line complete, with 32 to 34 scales; predorsal scales 14 or 15.

Colour and Size

Bright metallic-blue, head silvery, belly pale pink; three or four steel blue longitudinal bands along flanks, separated by narrow yellow lines. Fins yellow to deep orange-red; pectoral fins hyaline. It attains a maximum size of 10cm.

Food and Feeding

It is an omnivorous fish, readily accepts anything that its mouth can hold, not at all fussy about food. Prefers insect larvae.

Sexual Dimorphism and Breeding

Males develop intensive colouration during breeding. Females develop bulged belly. The breeding of danio is comparatively easier. It is a prolific breeder. The aquarium must be heavily stocked with plants preferably hydrilla. The sticky eggs adhere to the leaves. A pair consisting of single male and female is ideal for breeding. They must be well conditioned on live foods and introduced into the breeding tank, on the evening prior to the day on which breeding is expected to take place. If the male and female are conditioned in separate aquaria they should spawn the following morning. The optimum range of temperature is 22- 24°C. Parents are to be removed after spawning as they may devour the eggs and young ones. The young ones hatch out after 24 hours. On the third day onwards they feed on infusoria. After ten days they will start feeding on *Moina*. They need plenty of space for further growth.

Aquarium Requirements

Less demanding fish, but prefers clear and aerated water. It can thrive well even in non-aerated water. It is easily acclimatizable.

Behaviour in Captivity

It is a popular aquarium fish that moves in groups and remains in the water column. It is compatible and very active in aquariums.

This is an admirable fish and is considered as one of the standard members of a 'happy family'. It matures at 6-7cm. This gorgeously coloured fish is popular with aquarists. It is peaceful and well behaved in community aquaria, and its hardiness and readiness to breed further enhances its attractiveness.

Osteobrama bakeri (Day)

Common Name

Malabar osteobrama (English)

Fin Count

Dorsal iii 8; Anal iii 11; Pectoral i 12; Ventral i 10

Distinguishing Characters

Body trapezoid and considerably compressed, its depth about 3.1 times in standard length; abdominal edge sharp and trenchant between bases of pelvic and anal fins, but, rounded in front of pelvic fins. Mouth small; barbels two small but well-defined pairs. Dorsal spine weak and serrated. Scales small; lateral line with about 44 scales; scale-rows 5½ between lateral line and base of pelvic fins; predorsal scales 15.

Colour and Size

Silvery, dorsal profile with metallic blue colour, dorsal and caudal fins with orange coloured edges. Attains a maximum size of 12cm.

Food and Feeding

It is omnivorous.

Sexual Dimorphism and Breeding

Not known.

Aquarium Requirements

Difficult to acclimatize to captive conditions. Very sensitive and requires well-aerated clear water.

Behaviour in Captivity

It is a very compatible and peaceful fish. Remains at the middle part of the tank.

Endemic to the Western Ghats

Puntius amphibius **(Valenciennes)**

Common Name

Scarlet banded barb (English)

Fin Count

Dorsal ii-iii 8; Anal ii-iii 5; Pectoral i 14; Ventral i 8

Distinguishing Characters

Body spindle-shaped, both of its profiles equally convex, its depth about 3.6 times in standard length. Head length about 3.75 times in standard length. Mouth small and subterminal; barbels one pair of maxillary, shorter than orbit. Dorsal fin inserted nearer to tip of snout than to base of caudal fins, its last unbranched ray feebly osseous and smooth. Scales medium; lateral line complete, with 23 or 24 scales; predorsal scales 7 or 8.

Colour and Size

Upper half steel-blue, fading to white with golden tinge on flanks and abdomen; a large well-marked black spot on base of caudal fin; sexually mature fish with scarlet lateral band from eye to caudal fin. Dorsal fin orange with an oblique band; other fins yellowish.

Food and Feeding

It is omnivorous. They accept pellet feed in aquarium tanks. They prefer live feed like tubifex, earthworms or *Chironomus* larvae. They are slow feeders. Specific attention should be paid to note that they get food in a community aquarium.

Sexual Dimorphism and Breeding

When sexually mature, males develop a scarlet band along the mid lateral side of the body extending from behind the opercle to the tail region. Hence they are commonly called scarlet banded barbs. Females do not have this colour, but they have a bulged belly.

Aquarium Requirements

This species prefer to be at the bottom part of the tank. They require well-aerated water with neutral pH. Being shy initially, they need some hiding places in the tank for easy acclimatization.

Behaviour in Captivity

These are very compatible and peaceful species. Usually they remain towards the bottom part of the tank. Initially they are very shy. In a community tank they hide among the plants or behind the stones, but in a single species aquarium, where these fishes are kept alone, they dwell in all parts of the tank. However, they do not usually come to the surface of the water column.

Puntius arulius arulius (Jerdon)

Common Name

Aruli barb, Longfin barb (English)

Fin Count

Dorsal iii 8; Anal ii 5; Pectoral i 14; Ventral i 8

Distinguishing Characters

Body elongate and fairly compressed, its depth 3.2 to 3.5 times in standard length. Head 3.5 to 3.7 times in standard length. Mouth moderate; barbels one pair of very thin maxillary only. Dorsal fin inserted generally nearer to tip of snout than to base of caudal fin, often equidistant; its last unbranched ray non-osseous and fairly weak. Scales moderate; lateral line complete, with 21 to 24 scales; predorsal scales 8. A prominent, fairly deep pectoral pit present.

Colour and Size

Back olivaceous-green, blends to silvery on belly with a reddish lustre; scales over lateral line particularly with numerous tiny green shiny spots; operculum with an iridescent green dot; four or five

black blotches on body, the transverse bars at level of dorsal-fin origin, at level of anal fin and on caudal peduncle extremely prominent. Caudal fin yellowish to reddish, with bright red tips. It attains a maximum size of 18cm.

Food and Feeding

Active feeder, prefers insect larvae, readily accepts any food including pellet feed. Requires plant feed also.

Sexual Dimorphism and Breeding

During breeding males can be easily identified by the presence of long filamentous rays of dorsal fin, extending beyond the fin-membrane. It also develops intensive colouration on the edges of fins, caudal fin becomes red. Females do not have these characters but have a bulged belly.

Aquarium Requirements

Needs larger tanks and well aerated clear water. Not very easily acclimatisable. Cannot tolerate wide variation in pH, temperature or oxygen.

Behaviour in Captivity

Very active and usually dwells at the upper part of the water column. Males become aggressive when they are sexually mature. Two mature males in a tank are not compatible to each other. Otherwise, good for community aquarium.

Puntius denisonii (Day)

Common Name

Denison barb, Miss Kerala, Red-line Torpedo (English)

Fin Count

Dorsal ii-iii 8; Anal iii 5; Pectoral i 14; Ventral i 8

Distinguishing Characters

Body rather deep, its depth about 3.75 times in standard length. Head about 4.5 times in standard length. Mouth small; barbells one maxillary pair only, longer than orbit. Dorsal fin inserted nearer to tip of snout than to base of caudal fin; its last unbranched ray non-osseous, weak and articulated. Scales medium; lateral line complete, with about 28 scales; predorsal scales 9.

Colour and Size

Silvery with a black band, above which runs a horizontal fluorescent scarlet stripe passing from snout to base of dorsal fin. Caudal fin with oblique yellow and black bands crossing the posterior-hall of each lobe. Dorsal fin is anterior in position and with scarlet colouration. It attains a maximum size of 24 cm. Fishes from River Valapattanam, Iritty, Kerala are more colourful on their dorsal fins.

Food and Feeding

It accepts live insects, worms, mosquito larvae etc. Also accepts pellet feed. It is very shy and in a community tank special care should be taken to note that this fish gets food. Not an active feeder. In single species aquarium, it is very active.

Sexual Dimorphism and Breeding

No dimorphic characters are observed. Females, when they are mature, have slightly broader abdomen than that of the male. Mature male oozes out milt when pressed at the vent region. Not bred so far in captivity.

Aquarium Requirements

Needs clear and well-aerated water.

Behaviour in Captivity

Peaceful and compatible, very shy. It is a very sensitive fish, difficult to acclimatize in captive condition.

One of the prettiest barbs, and exhibits stunning colouration. Endemic to Kerala part of the Western Ghats. The species has won the title "Most Attractive Fish" in several aquashows both in India and abroad. Popularly named as "Ms. Kerala". Taxonomic identity of similarly looking *P. chalakkudiensies* needs further confirmation using molecular markers and work on this aspect is being carried out at NBFGR lab, Kochi.

Puntius filamentosus (Valenciennes)

Common Name

Black-spot barb, Filament barb (English)

Fin Count

Dorsal iii (iv) 8; Anal ii-iii 5; Pectoral i 14; Ventral i 8

Distinguishing Characters

Body elongate, its depth 3.3 to 3.8 times in standard length. Head about 4.3 times in standard length. Mouth moderate; barbels a very small pair of maxillary only, often hidden in grooves round the corners of mouth. Dorsal fin inserted equidistant between tip of snout and base of caudal fin, its last unbranched ray non- osseous, weak and smooth; in adult males, generally five unbranched rays

elongated into filaments. Scales large; lateral line complete, with about 21 scales. Adult males studded with large tubercles on snout.

Colour and Size

A beautiful fish showing various colour patterns at different stages of life. Adults uniformly silvery to greenish-silvery, somewhat darker (olive-coloured) above, with a full rainbow sheen by reflected light; a distinctive dark oval blotch on lateral-line. Fins delicate yellow-greenish; dorsal fin rays partly dark violet, often dark tipped. Juveniles silvery, with broad deep black vertical stripes, and orange-red to brick-red fins; caudal fin of half-grown fishes reddish with a black blotch on each lobe, its tips whitish. It attains a maximum size of 22cm. One of the largest barbs of India.

Food and Feeding

Accepts all kinds of food including pelleted food. The surface is watched constantly and they feed from the upper column of water hence floating feed is preferable. Usually do not feed from the bottom.

Sexual Dimorphism and Breeding

Males have first few rays of the dorsal fin elongated when mature. The snout is covered with a patch of large tubercles on either side in front of the eyes. They are also more brightly coloured during spawning period. Females have a bulged belly when fully mature. It could be successfully bred in captivity by the first author. It breeds in ponds or in very large aquaria, heavily laden with plants.

Aquarium Requirements

It is a hardy species. But require aerated clear water. The male needs plenty of elbowroom. Then only they develop their full beauty. A top lighted tank is preferable. The fish is comfortable in a subdued, indirect light.

Behaviour in Captivity

It is a compatible species, very active and moves along all the parts of the water column. Young ones are very beautiful with the vertical bands on their body. The tip of tail with black and red colour adds more to its beauty. Since it grows to a fairly big size, this is also considered as a food fish. This can be recommended as very good garden fish also.

Puntius jerdoni (Day)

Distinguishing Characters

Body fairly deep, its depth 2.7 to 3 times in standard length. Head 4.2 to 5 times in standard length. Mouth narrow; barbels two pairs, maxillary pair equal to orbit, rostral pair slightly shorter. Dorsal fin inserted equidistant between tip of snout and base of caudal fin; its last unbranched ray non-osseous, weak and articulated. Scales medium; lateral line complete, with 26 to 32 scales; predorsal scales 12.

Colour and Size

Milky white or glittering silvery body. Fins with fluorescent orange, tipped with black. It attains a maximum size of 46cm.

Food and Feeding

This fish accepts all kinds of food including artificial feed. Prefers live food.

Sexual Dimorphism and Breeding

The female has a smooth snout while the male has tubercles on the snout. No other dimorphic characters could be observed yet to be bred in captivity.

Puntius sarana subnasutus (Valenciennes)

Common Name

Peninsular olive-barb (English)

Fin Count

Dorsal iii 8; Anal ii 5; Pectoral i 16; Ventral i 7

Distinguishing Characters

Body oblong and fairly deep, its depth 2.7 to 2.9 times in standard length. Head fairly small, its length 4.4 to 4.8 times in standard length. Eyes moderate, its diameter about 3.5 times in head length. Mouth moderate; barbels two pairs, maxillary pair much longer than orbit, rostral pair slightly shorter. Dorsal fin inserted equidistant between tip of snout and base of caudal fin; its last unbranched ray osseous, fairly strong weak in young and posteriorly serrated. Scales moderate; lateral line complete, with 28 to 31 scales; predorsal scales 10.

Colour and Size

Silvery on back and upper half of body, fading to white with gold beneath; most scales with black bases; a dark band behind operculum and a black blotch on lateral line on about the 24th scale. Fins orange; caudal fin with a black superior and inferior edge. It attains a maximum size of 30 cm. Juveniles ornamental; a cultivable food fish owing to fast growth rate.

Food and Feeding

It feeds on both natural and artificial food. Not at all fussy about type of food.

Sexual Dimorphism and Breeding

It can be bred in captivity. The breeding and larval rearing technology has been developed

Aquarium Requirements

It is a hardy species. But prefers clear and aerated water. It can thrive well even in non - aerated water.

Behaviour in Captivity

It is a very compatible and peaceful fish, good for community aquariums. The juveniles are very beautiful as ornamental fish.

Osteochilus (*Osteochilichthys*) *nashii* (Day)

Common Name

Nash's barb (English)

Fin Count

Dorsal iv 11; Anal iii 5-6; Pectoral i 14; Ventral i 8

Distinguishing Characters

Body oblong and compressed, its depth 3.2 to 3.6 times in standard length. Snout overhanging mouth, in adults covered by papillae. Mouth broad and inferior; jaws in young compressed, each with a cartilaginous covering, with growth the mouth widens and cartilaginous covering becomes more horny; lips simple and continuous at angles of mouth, lower lip between lateral portions of labial groove considerably behind tip of lower jaw to which it is firmly attached and is plicated; barbels absent. Dorsal fin without any osseous ray. Scales moderate-size; lateral line with 40 to 43 scales.

Colour and Size

Reddish-brown along back, abdomen silvery; a black lateral band from eye to caudal fin. Dorsal fin with a dark band on middle, the band edged above with scarlet; a dark band on anal fin. Young silvery grey on back, fading to silvery on sides; lateral band terminates in a dusky blotch at base of caudal fin. Attains a maximum size of 18cm.

Food and Feeding

It is omnivorous in habit but prefers live food. Readily accepts any food in captive conditions.

Sexual Dimorphism and Breeding

Not known.

Aquarium Requirements

Needs well aerated clear water. It is hardy and easily gets acclimatized. Active swimmer in aquarium.

Behaviour in Captivity

Compatible and peaceful fish. Usually dwells towards the bottom part of the water column.

Garra mullya (Sykes)

Common Name

Mullya garra (English)

Fin Count

Dorsal iii 7-8; Anal I-ii 5; Pectoral i 12-15; Ventral i 7-8

Distinguishing Characters

Body slightly flattened, its depth 3.8 to 4.3 times in standard length. Head somewhat flattened on under-surface; mouth rounded and smooth, with the tip marked off by a deep transverse groove; interorbital region somewhat convex, its width 1.8 to 2.3 times in head length. Mouth small; suctorial disc small but well-marked, its width 1.5 to 2.2 times in head width. Barbels two pairs; rostral pair as long as or slightly shorter than eye-diameter, maxillary pair shorter than rostral ones. Dorsal fin inserted nearer tip of snout than to caudal fin base. Pectoral fins shorter than head length. Caudal fin slightly emarginate. Scales moderate-size; lateral line with 32 to 34 scales; lateral transverse scale-rows 4½/3½; predorsal scales 9 to 11; breast and belly often naked. Distance of vent from anal fin 3.6 to 3.8 times in inter-distance between pelvic fin origin and anal fin base.

Colour and Size

Upper surface of head and body, and flanks darkish; a broad lateral band on sides, bordered above and below by incomplete dark narrow lateral stripes especially in posterior half of the body; belly dull white; a distinct black spot just behind the angle of operculum; a dusky blotch at caudal fin base. It attains a maximum size of 17cm.

Food and Feeding

It is an omnivorous fish. Since they are mainly vegetarians they avidly devour artificial feed. It is an algae eater. Its fantastically large ventral suctorial lips are well adapted to scrounging around for algae. Because of its algae eating habit it is a good glass cleaner. It browses algae from the glass surface.

Sexual Dimorphism and Breeding

No clear cut sexual dimorphism is observed, but females can be identified by the bulged belly when mature. Male oozes out milt on gentle press on the belly. Captive breeding is possible by hormonal application.

Aquarium Requirements

Needs well-aerated water. It can be easily acclimatized to captivity.

Behaviour in Captivity

It is a bottom feeder. But it is quite interesting to watch it browsing algae from the glass surface, drift wood kept in tanks and leaves without damaging them.

Acanthocobitis moreh (Sykes)

Common Name

Moreh loach (English)

Fin Count

Dorsal iii 9-10; Anal iii 5; Pectoral i 11; Ventral i 7

Distinguishing Characters

Body spindle-shaped, its depth 4 to 4.1 times in standard length. Eyes large, not visible from underside of head. Nostrils close to each

other. Mouth semicircular; lips fleshy, upper lip with a few rows of papillae, lower lip interrupted in middle with two, rounded, raised clusters of small papillae situated on each side of cleft of lower jaw. Barbels well developed; nasal pair short. Dorsal fin inserted nearer to snout-tip than base of caudal fin. Caudal fin slightly emarginate. Lateral line incomplete, ending opposite to posterior end of dorsal fin.

Colour and Size

Body marked with several broad black bands and spots. Fins are also marked with dark streaks on their rays. Attains a length of 4.4 cm.

Food and Feeding

It is omnivorous, but prefers insect larvae. Accepts artificial feed in captivity.

Sexual Dimorphism and Breeding

Not known.

Aquarium Requirements

Needs well aerated clear water. Requires hiding places for comfortable settlement in the tank.

Behaviour in Aquarium

Remains towards the bottom part of the tank. Compatible and peaceful.

Nemacheilus monilis (Hora)

Common Name

Moniliform loach (English)

Fin Count

Dorsal iii 7; Anal ii 5; Pectoral i 10; Ventral i 6-7

Distinguishing Characters

Body rather elongate and of uniform depth, its depth about eight times in standard length. Eyes small, not visible from underside of head. Nostrils close to each other; anterior nostrils somewhat tubular. Mouth semicircular; lips moderately fleshy and poorly furrowed, upper lip raised into a short proboscis in middle, lower lip interrupted in middle. Barbels relatively long and thread-like.

Dorsal fin inserted slightly nearer base of caudal fin than to snout-tip. Pelvic fins separated from anal-opening by a considerable distance. Caudal fin deeply forked, with pointed lobes. Scales small and imbricate, indistinct in anterior part of body, absent on ventral surface; lateral line complete. Vent situated some distance in front of anal fin.

Colour and Size

Dirty white and somewhat infuscated along back; a distinct moniliform black band along lateral line from tip of snout to base of caudal fin, interrupted by eye in its course; the last component of black spots in the series more prominent; the band continued as a black streak in middle of caudal fin; barbels streaked with black. Fins whitish. It attains a maximum size of 8cm.

Food and Feeding

Omnivorous in feeding habit. They relish mosquito larvae, boiled egg yolk and artificial feed. Usually they feed from the bottom, but if they are too hungry they swim up to the upper parts of water column when food is provided. If they are too hungry they do not wait for the food to reach the bottom. Recommended food is sinking pellets, earthworms, live food like mosquito larva, moina and boiled chicken egg yolk.

Sexual Dimorphism and Breeding

Females develop bulged belly when they are mature. Breeding techniques developed for *Nemacheilus triangularis* can be applied for this species also.

Aquarium Requirements

They are more active than other laoches. Usually roam around the tank in search of food.

Behaviour in Captivity

They always dwell at the bottom of the tank. They are compatible and peaceful, and it is quite interesting to note that they come out of the hiding places when food is given. They are playful and friendly fish, prefer to hide and will be comfortable only when hiding places are provided in the tank.

Schistura denisoni denisoni **(Hora)**

Common Name

Denisonii loach (English)

Fin Count

Dorsal iii 8; Anal ii 5; Pectoral i 10; Ventral i 6

Distinguishing Characters

Body of uniform depth from 5.3 to 7.3 times in standard length. Eyes small, not visible from underside of head. Nostrils close to each other, anterior tubular. Mouth semicircular; lips fleshy, lower weakly furrowed and interrupted. Barbels well-developed, thread-like, as long as eyediameter. Dorsal fin inserted equidistant between snout-tip and base of caudal fin, often slightly nearer caudal fin base. Pelvic fins touching anal-opening or close to it. Caudal fin deeply emarginate. Scales small, imbricate posteriorly, close-set in the middle, scattered interiorly, absent on the under surface; lateral line incomplete, ending at origin to middle of dorsal fin.

Colour and Size

Body marked with 12 or 13 broad vertical bands with an equal number of narrow pale interspaces; a black band at base of caudal fin; and a blackish spot at origin of dorsal fin base. Dorsal fin with two rows of spots; caudal fin with four rows of well-marked spots; other fins unspotted. Attains a maximum size of 10cm.

Food and Feeding

Omnivorous in feeding habit. It prefers live food, but readily accepts artificial feed also.

Sexual Dimorphism and Breeding

Male has a small hook-like structure on its head. Female develops bulged belly when becomes mature. Male and female can be easily identified by their body shape, when they are mature.

Aquarium Requirements

Needs well aerated water. They are comfortable in the tank only when some hiding places are provided.

Behaviour in Captivity

Peaceful and compatible. Always dwells at the bottom of the tank, hiding among pebbles, drift wood or under big stones. Two sub-species, *S.denisoni mukambbikaensis* (Menon) and *S.denisoni pambaensis* (Menon) have also been reported from River Kollur in Karnataka and River Pampa in Kerala respectively.

Mesonemacheilus guentheri (Day)

Common Name

Guenther's loach (English)

Fin Count

Dorsal ii 8; Anal ii 5; Pectoral i 10; Ventral i 6-7

Distinguishing Characters

Body somewhat subcylindrical with head and anterior part of body fairly flattened. Eyes small, not visible from underside of head. Nostrils close to each other; anterior nostrils slightly tubular. Mouth semicircular; lips fleshy and deeply furrowed, lower lip interrupted in middle. Barbels well developed. Dorsal fin inserted equidistant between tip of snout and caudal fin base. Caudal fin forked. Scales small, imbricate in posterior-third of body, isolated anteriorly in front of dorsal fin, absent on ventral side of body; lateral line almost complete, ending above anal fin.

Colour and Size

Dark brown with three rows of whitish spots of different sizes and form; a deep short vertical bar at base of caudal fin; a spot on dorsal fin origin. Dorsal and caudal fins with three or four rows of spots. Attains a length of 6 cm.

Food and Feeding

It is omnivorous in feeding habit.

Sexual Dimorphism and Breeding

Males and females can be identified when they are sexually mature. Females develop bulged belly. Males develop red colour on their caudal fin on attaining maturity.

Aquarium Requirements

Needs clear and well-aerated water. As in the case of other loaches they also remain towards the bottom of tanks. They are comfortable only when hiding places are provided.

Mesonemacheilus triangularis (Day)

Common Name

Triangularis loach (English)

Fin Count

Dorsal ii 8; Anal ii 5; Pectoral i 10; Ventral i 7

Distinguishing Characters

Body subcylindrical, its depth 5 to 8.5 times in standard length. Eyes small, not visible from underside of head. Nostrils separated by a prominent flap. Mouth semicircular; lips fleshy and furrowed, lower interrupted in middle. Barbels well developed. Dorsal fin inserted about equidistant between snout-tip and base of caudal fin. Caudal fin forked. Scales distinct, imbricate all over body; lateral line most complete, reaching up to caudal peduncle. Vent a short distance from anal fin.

Colour and Size

Varies considerably with age; ground colour grayish with yellowish bands edged with black on body, usually seven, the anterior five directed obliquely backwards, and last two vertical; several yellowish patches of different patterns above lateral line; a dusky blotch on base of caudal fin. Dorsal and caudal fins with two bands each; and anal and pelvic fins with one each. It attains a maximum size of 8 cm.

Food and Feeding

They eat anything like worms, flakes or moina. They also enjoy mosquito larvae, boiled egg yolk and artificial feed. They feed from the bottom, hence care should be taken to note that the feed reaches the bottom in a community tank.

Sexual Dimorphism and Breeding

Females develop bulged belly when they are mature. The first author could successfully develop captive breeding technique for this species. Young ones become mature when they are one year old.

Aquarium Requirements

They become comfortable only when provided with convenient hiding places. To make them comfortable we provided them with pieces of PVC pipes in the tank bottom, so that they would do better in the tanks.

Behaviour in Captivity

They always dwell at the bottom of the tank. Never come to the top layers of water column. They are compatible and peaceful, and it is quite interesting to note that they come out of the hiding places when food is given. They seem to be doing better in groups, though they do not spend all the time together. This species easily get scared by any movement.

Lepidocephalus thermalis (Valenciennes)

Common Name

Malabar loach (English)

Fin Count

Dorsal ii-iii 5; Anal ii-iii 5; Pectoral i 6-7; Ventral i 6

Distinguishing Characters

Body elongate, low, slightly compressed anteriorly and strong posteriorly, its depth 7.5 to 9.7 times in total length. Mouth inferior; barbels three pairs; mental lobe well-developed, with small barbel like prolongations. Dorsal fin inserted somewhat anterior to pelvic fins, usually nearer caudal fin base than to snout-tip. Caudal fin almost squarely truncate. Scales very small; a small patch of scales on head behind suborbital spine; on ventral side of head scales extent anterior to pectoral fin bases but not reaching isthmus; 30 to 37 rows of scales between back of body and anal fin; scales oval.

Colour and Size

Grey to delicate grey-green, with somewhat dark 8 to 10 irregular blotches along flanks; back usually marbled with pale and dark; a small black spot on upper half of base of caudal fin dorsal and anal fins with rows of spots. An attractive ornamental fish. It attains a maximum size of 8cm.

Food and Feeding

Wide variety of food is accepted, especially sinking dried food and bottom live foods.

Sexual Dimorphism and Breeding

Not bred under captivity so far.

Aquarium Requirements

Water chemistry is not critical and wide range of pH and water hardness will be tolerated without distress. Temperature should be 24–26°C.

Behaviour in Captivity

Peaceful, good for community tanks with similar sized tankmates. They enjoy digging and burrowing into the sand. Hence a rocky substrate tends to wear down the delicate barbels. So a sandy substrate is preferable. They should have plenty of hiding places, they tend to stuck in small openings, so should be very careful.

Scatophagus argus (Linnaeus)

Common Name

Scat (English)

Fin Count

Dorsal xi 16-18; Anal vi 14-15; Pectoral 16-17; Ventral i 5

Distinguishing Characters

Body quadrangular, strongly compressed. Head profile rising steeply to nape; snout and interorbital space rounded. Mouth small, with brush-like teeth. Dorsal fin deeply notched; dorsal fin membrane incised between spines. Scales very small.

Colour and Size

Variable colour patterns. Young fishes of about 2 cm are usually quite dark in colour; finest colouration and markings are attained in fishes of about 5-6 cm total length; uniform greenish-silvery, bluish-silver or coffeebrown with a delicate golden- sheen, especially on back; numerous dark spots mainly confined to upper portion of sides. Attains a maximum size of 30cm. Widely distributed in India.

Food and Feeding

Omnivorous, but accepts pelletted feed also.

Sexual Dimorphism and Breeding

Breeding has been standardised in CIBA, Chennai.

Aquarium Requirements

It is euryhaline and can withstand salinity up to 25ppt. It can be gradually acclimatized to freshwater conditions. It should be kept in large tanks with plenty of water.

Behaviour in Captivity

Hardy and compatible, ideal species for aquarium.

Remarks

An attractive fish with different colouration when they are young. Some fishes have red spots arranged in various patterns on the back; such form are known as *Scatophagus rubrifrons*, a name which is quite incorrectly given since these are merely varieties of *S. argus* and should at the most be called 'Red Argus'.

Tetraodon travancoricus (Hora and Nair)

Common Name

Malabar Puffer Fish (English)

Fin Count

Dorsal 7-8; Anal 8; Pectoral 16-17

Distinguishing Characters

Body oblong and compressed laterally; dorsal profile arched, highest at midst of back; interorbital space flat. Mouth terminal, directed forward. Nostril hollow tube, nearly as high as wide, only its distalmost part formed into two very small lobes which are bent inwards giving a key-hole appearance to opening of nasal organ. Body spinules inconspicuous.

Colour and Size

Colour of upper parts grayish, of lower parts much lighter; usually two black, oval patches on upper lateral surface of body in front of dorsal fin; posterior to the patches, a dark, broad band running to caudal fin and partly continuing to central rays; usually a dark spot in middle of course of band; other dark spots at base of caudal fin and at base of posterior most two dorsal fin rays; a dark patch above pectoral fin and a spot behind it; dorsal surface with a narrow, light interocular band, two irregular dark patches behind eyes and followed by a V-shaped marking; an irregular band in front of dorsal fin and triangular patches in front of or behind dorsal fin. Fins hyaline. Attains a maximum length of 3cm and it is the smallest species in the family.

Food and Feeding

Feed only on live food. They hardly take artificial feed. It is helpful in removing molluscs like *Planorbis* in aquarium tanks. They love to eat snails. This can be a curse or a blessing. If you have an unwanted snail outbreak from new plants, a few puffer fishes will help crush the infestation. On the other side it is not advisable to keep decorative snails like apple snails or *Planorbis* in a tank with puffer fish.

Breeding Requirement

Not known.

Aquarium Requirements

Very easy to acclimatize to captive conditions. They require clean well-aerated water.

Behaviour in Captivity

Compatible and peaceful. But they are notorious for nipping the fins of other fishes. The fins thus injured will be infected and may become lethal to the fish. They always remain towards the upper half of the water column. They are slow-moving and excellent candidates for monospecies aquariums.

Pristolepis fasciata (Bleeker)

Distinguishing Characters

Body thick-set, very deep and compressed. Mouth moderate; teeth villiform on jaws and palatines. Eyes moderate sized. Dorsal spines strong. Lateral line interrupted opposite posterior end of dorsal fin, continued on third row of scales below it, with 26-28 scales.

Colour and Size

Greenish or brownish yellow with a deep black spot on shoulder and another over upper part of pectoral fin base. Fins with reddish edge.

Food and Feeding

Omnivorous. But prefers live food.

Sexual Dimorphism and Breeding

Males and females can be identified by examining the genital papilla.

Aquarium Requirements

Survives in clean aerated water.

Behaviour in Captivity

Peaceful and compatible fish, well suited for community quariums. It is a delicious food fish. Very common in brackishwater of the coastal regions of Kerala. To analyse the taxonomic relationship of this species from Kerala with the same from Mekong River, Myanmar and Southeast Asia, further investigations are required.

Prisolepis marginata Jerdon

Common Name

Malabar Catopra (English)

Fin Count

Dorsal xiv-xvi 11-14; Anal iii (rarely iv) 8; Pectoral 14-15; Ventral i 5

Distinguishing Characters

Body oblong and compressed. Mouth moderate; teeth villiform on jaws, outer row of teeth somewhat enlarged and in some specimens two or four only enlarged in lower jaw; teeth villiform on vomer. Dorsal spines rather stout; second anal spine strongest but as long as third spine. Lateral line interrupted (divided) opposite fourth dorsal fin ray on 21st scale, with 25 to 27 scales.

Colour and Size

Greyish-green with purplish reflections; often vertically banded. Fins with lighter edges; caudal fin with whitish out edge. Attains a maximum size of 17cm.

Food and Feeding

Omnivorous, but accepts pellet feed also.

Sexual Dimorphism and Breeding

Exhibits no sexual dimorphism except during the time of breeding. Male has an enlarged anal papilla and female with a pot belly. But sexes can be identified by the behaviour in captivity. Mature male exhibits territorial behaviour. Captive breeding and larval rearing techniques have been developed. This fish has some interesting spawning behaviour that a hobbyist should note. They are guarders and lithophil spawners. The mature male fish starts nest building by selecting a portion in the tank and displaying territorial behaviors. The male exhibits greater aggressiveness and territoriality and is busily engaged in preparing a nest in the pebbly bottom. The nest of common catopra is nothing but a small depression like-structure made of pebbles. For the preparation of pebble nest the male carries big-sized pebbles to the proposed site of the nest and at the same time takes away smaller ones and sand particles. Their thick lips are the only organs used as tools. Each stone is carefully fanned for cleaning after placing it in the pit. The completed nest is in the form of a pit with little thick risen border. The nest is very clearly distinguishable from other parts of the tank. They prefer to have their honeymoon in the pebbly bottom. When the pit is ready, the female is invited to the nest. The courtship rituals include the sidewise lying inside the pit, along with shivering of the fins and body. They circle inside the pit and shake their fins and body vigorously. They incline slightly to one side and keeping the anal region close to one another, the female with a more enlarged genital papillae releases a few eggs, the milt from the male simultaneously fertilizes the eggs. The eggs fall into the voids of the stones. After a short while, the same act is repeated several times. The female tries to devour the eggs whenever it gets a chance but the male defends her. When the spawning activities are completed, the male starts guarding the eggs by fanning with its fins and defending any intruders. The male also continues to rearrange the pebbles until the pit is changed into a heap. After four days the free swimming larvae comes out through the gaps in the pebbles. The fish spawns year round. The success of this research indicates the possibility of commercial production of this rare species for ornamental trade.

Aquarium Requirements

Requires gravel bottom, survives well in aerated clean water.

List of Other Indigenous Ornamental Fishes Recorded from Western Ghats

N. monilis

Garra surendranadhini

P. malabaricus

Bhavania australis

Garrah Cleaner

Nemacheilus guentheri

Melanbarb-male

Nemacheilus deniusonii

Behaviour in Captivity

Though not brightly coloured this grayish-green coloured species makes an excellent candidate for your aquarium. Its behaviour is pretty enough to catch your eyes and challenging enough to keep your interest. It is a slow moving fish, standing most of the time in still position, moving its pectorals. It readily accepts any food under captive conditions and is compatible with other inmates in the aquarium except during its breeding time.

Other new species of importance are *Puntius exclamatio* and *Puntius pookodensis* both are endemic to the Western Ghats.

References

Anna Mercy, Gopalakrishnan, Kapoor and Lakra, 2010. *Ornamental Fishes of Western Ghats, India*. National Bureau of Fish Genetic Resources, Lucknow Publication.

Chapter 4

Captive Breeding and Conservation Strategies for Indian Freshwater Ornamental Fish

Saroj K. Swain[1], *N. Rajesh*[1] *and Ambekar E. Eknath*[2]

[1]*Central Institute of Freshwater Aquaculture (CIFA), (Indian Council of Agricultural Research) Kausalyaganga, Bhubaneswar – 751 002*

[2]*Director General, Network of Aquaculture Centres in Asia-Pacific (NACA) PO Box 1040, Kasetsart Post Office, Bangkok 10903, Thailand*

Ornamental fish play an important role in interior decoration and its culture for aquarium trade is a fast expanding industry. The in-door aquarium keeping is the second most popular hobby, after photography, among people all over the world due to their attractive coloration, aesthetic beauty and the pleasure and soothing effect they give to the eyes and help in relieving stress. Farming and trading of ornamental fish has started to gain considerable momentum in India. The focus, however, has been on formulation of green protocol that will ensure conservation of fish diversity and the environment and at the same time contribute to the well-being of local fishermen and fair trade. The major concern is to conserve some of the important species available naturally. Central Institute of Freshwater Aquaculture (CIFA) has been implementing a focused program on conservation and captive breeding of important indigenous ornamental fish. The general approach has been to gather information on the habitat, ecology, feeding and reproductive behaviour of the target species and to develop captive breeding techniques. This knowledge is being disseminated to the users and various other stakeholders for developing a sustainable ornamental fish trade in the country.

There is also a craze for keeping new varieties of fishes either they are from bred varieties or collected from the nature. Therefore new fish varieties are seen from time to time in the market. By concentrating on such fishes, we may loose our indigenous stocks for which many organizations in the world are now spending millions of dollars towards preservation of the indigenous endangered stocks as germplasm. Indigenous ornamental fishes could be useful for developing new strains to compete in world market. Most of the ornamental fishes produced all over the country are channelled to Singapore and exported all over the world. Singapore, the world's largest exporter of ornamental fish with 30 per cent of the global supply, has already produced numerous lucrative new varieties through selective breeding. Of total exports of ornamental fishes freshwater fish accounted for around 90 per cent, with almost 90 per cent of them bred in captivity. The total worth of ornamental fish trade is 280 million US Dollar.

Global Trade of Ornamental Fishes

The global ornamental fish trade has been increasing at a steady pace since 1980's reaching US$ 4.5 billion in 1995, which further has been striding at an impressive annual growth rate of about 14 per cent. In the year 2007 the marine component of the trade has grown to 48 per cent and the freshwater component to 52 per cent. According to FAO, (2009) the value of brackish-water fish traded annually is negligible. The international ornamental fish trade in retail level is estimated to be more than US$ 8 billion, while the entire industry includes accessories, aquarium, feed, medicines, plants etc. is estimated to the tune of US$ 20 billion. During 2008 more than 125 countries were involved in the ornamental fish trade, out of which 15 countries exported fish worth more than US$ 5 billion, among which seven are Asian countries. Asian countries together contributed more than 52 per cent of the global export.

All together 1800 species of fish are traded globally of which majority are from marine environment including crustaceans and invertebrates. Among freshwater fish species there are 35 species dominate the whole market. The species generally dominating the market are livebearers (guppy, platy, swordtail and molly) and tetra (neon tetra). The average retail price for food fish was between US$ 14,500 and US$ 16,500/t, but ornamental fish typically trade at a US$ 1.8 million/t.

Asian Trade

During the year 2008, 31 countries exported ornamental fishes with top ten countries from Asia with over US$ 70.4 million. Singapore remains in top of the list among Asian trade with over 400 species and 1000 varieties of ornamental fish. Singapore is considered as the super market for ornamental fish. Malaysia is the second largest global exporter and second largest Asian supplier.

Indian Irnamental Fish Resource and its Contribution in World Trade

Indian water possesses a rich biodiversity of about 2118 species, out of which more than 600 species exist in freshwater. The country further possesses a rich diversity of colourful ornamental fish, with over 100 varieties of indigenous species, in addition to the same number of exotic species that are bred in captivity. It has also been noticed that Indian ornamental fishes are of greater demand in international market. The main markets are the US, the UK, Belgium, Italy, Japan, China, Australia and South Africa. With its tropical climate, India can become a key player. Many Indian species like catfish, dwarf gouramis and barbs are popular abroad and fetch good prices.

India's share to the global export market is almost insignificant. At present the production of indigenous ornamental fishes from the country is mainly confined to freshwater varieties and is limited to fishes collected from nature, predominantly in the Northeastern (NE) states and the Western Ghats (85 per cent) and a few bred varieties of exotic species (15 per cent). According to an estimate there are 350 fulltime and more than 2000 part time fish breeders are available in the country. More than 2,500 species are traded and only 30 to 35 species of fresh water fish dominate the market. Kolkata has been the gateway for export of popular aquarium fishes from different part of country, from where they are distributed to other countries.

Hobby and Indian Domestic Trade

There are more than 1.5 million people keeping ornamental fish in their houses in the country. Further, it has been estimated that the domestic market in India is very large and presently, more than INR 100 crores of worth fish and aquarium related material is being traded annually in the country. There are more than 1,500 commercial outlets in the country involving retail trade of ornamental fishes and accessories. It has been observed that most of the aquarium breeding centers and the related trade has been confined in and around the metropolitan cities like Kolkata, Chennai and Mumbai, obviously due to ready urban market and the availability of international airport for both import and export business.

Biodiversity of Indigenous Species in India

As it has been evident from various research findings and deliberations in national and international meets for the indigenous fish varieties of India are dwindling day-by-day. For instance, the two pockets of North-Eastern Hill region and the Western Ghats, are still untapped. Out of 806 fish species inhabiting in freshwaters of India, the total fish species reported from northeast are 267, consisting of 114 genera under 38 families and 10 orders, which comprises, Assam-217, Arunachal pradesh-167, Meghalaya-165, Tripura-134, Manipur-121, Nagaland-68, Mizoram-48 and Sikkim-29, some of them being commonly found at more than one places (Personal communication, NERC, CIFRI, Guwahati, ASSAM) out of which 196 species are considered as ornamental. At present 58 species of indigenous ornamental fishes occurring in the northeast are currently being exported as reported by the MPEDA, India. Similarly, in Western Ghats out of 287 species, 192 species are demarcated as endemic.

As the ornamental fish diversity has scattered throughout the country and it is estimated by CIFA that, about 33 per cent are available in North-Eastern region, 24 per cent in Southern region, 23 per cent in Eastern region, 6 per cent in Western region, 3 per cent

Northern region, 2 per cent in Central region and 10 per cent from other regions of India.

Sustainable Harvest of Wild Species

- Freshwater fish can be sustainably harvested as long as natural habitats (including the breeding grounds) are conserved. For further works, knowledge of each of these important wild species of commercial and economic significance, a thorough knowledge about their habitat, ecology, food and feeding habit and feeding behaviour, breeding season and size, reproductive physiology (including fecundity) etc. need to be gathered by carrying out systematic studies.
- Annually, two phases occur in the 'wild – growing' and 'collapse' phases. Typically, the growing phase is just after spawning during the monsoon season and young ones or larvae are available in plenty. This phase is vulnerable to both inter- and intra-specific competition and predation. During growing phase certain species can be harvested.
- During the collapse phase many individuals die due to lack of water and other reasons. This is a particularly an appropriate time to harvest, not only because many of the fish will die anyway, but also it is relatively easier for fishermen to catch the fish in the shallow waters.
- Most species of tropical freshwater fish are not annual fish, and the typical lifespan of the smaller fish is from two to four years. Therefore, fish collected during the collapse phase, would be helpful for trade and hobbyists, and not have much impact on the population of the species.
- Proper harvesting techniques will certainly reduce the high mortality rates and the threat to the survival of the species can be avoided.
- Care must be taken to minimize the mortality in wild-caught fish as the fishermen use the catching device meant for food fish. The actual catching process needs to be improved.
- With improved catching methods and proper acclimatization before export will certainly enhance the quality and ensure a high percentage of survival on arrival.

- Improvement in packing technology, which allows high stocking densities for longer periods. This involves not only technology and water quality but also the knowledge of using appropriate anaesthetics and chemicals.

Development of Breeding Technology for the Indigenous Species at CIFA

The Central Institute of Freshwater Aquaculture (CIFA) is a premier research Institute in the country, working in R&D of different aspects of freshwater aquaculture. Ornamental fish breeding and its nutrition research is one of the recent origins. Special efforts are being made to breed some of the indigenous varieties in captive condition. Success in breeding and larval rearing has been achieved for a dozen ornamental fishes available in India (Table-1). A biotope based public aquarium has been established at Institute campus for education, research and conservation of Indigenous ornamental fish diversity besides many commercial fish species. Breeding and larval rearing of *Ompak pabda* is successfully done by the Kalyani field centre of CIFA leading towards commercialization.

Fishes Breed in Captivity

Basics to Undertake Breeding

- They should be well acclimatised before breeding in captivity.
- Male and female separation is required, 1-2 weeks prior to breeding as males can be identified from females because of the thinner abdomen compared to females.
- Feeding with nutritious food, especially bloodworms, brine shrimps or any hygienic live feed is essential.
- Water quality has to be kept at its optimum during the pre and post breeding conditions.
- After the conditioning period, the males and females are kept together in a same tank either in a perforated net cage or planted tank as per the need of the species.
- After putting both parents in the same tank the temperature may be enhanced to 27°C for easy spawning.

Table 4.1: Fish Species and the Types of Eggs

Fish Species	*Breeding Behaviour/Egg Type*
Brachydanio rerio (Zebra fish)	Egg scatterer, non sticky eggs
Brachydanio frankei (Leopard danio)	Egg scatterer, non sticky eggs
Puntius conchonius (Rosy Barb)	Egg scatterer, sticky eggs
Colisa fasciata (Banded gourami)	Nest builder, sticky eggs
Colisa lalia (Dwarf Gourami)	Nest builder, sticky eggs
Parlucioma daniconius (Black line rasbora)	Egg depositor, slightly sticky
Esomus barbatus (Flying barb)	Egg depositor, slightly sticky
Danio aequipinnatus (Giant danio)	Egg scatterer, non-sticky eggs
Danio devario (Torquoise danio)	Egg scatterer, semi sticky eggs
Puntius sophore (Sophore barb)	Egg scatterer, sticky eggs
Puntius ticto (Two spot barb)	Egg scatterer, sticky eggs
Badis badis (Chameleon fish)	Egg depositor, sticky eggs
Puntius fasciatus (Melon barb)	Egg scatterer, sticky eggs
Puntius filamentosus (Filament barb)	Egg scatterer, non sticky eggs

Breeding Techniques for Ornamental Fishes

Brachydanio rerio (Zebra Fish)

A small sized fish with a base colour of silver, marked by a number of bright blue lines which runs from the head to the rear edge of the caudal fin. The same pattern is repeated in caudal and anal fin too. The dorsal area is yellowish olive. This is a very active species, constantly swimming in upper strata of the aquarium. This can be bred in large numbers by photoperiod method in aquarium or concrete tanks. The bottom of the tank may be covered with marbles. The water height may be kept at least 10-15 cm for this fish to breed. The next day, by morning the danio parents are to be removed. Spawning generally takes place early in the morning and the non-sticky eggs are scattered in between the marbles so that the parents could not eat them. Fecundity ranges from 250-300 eggs/ female. Egg hatch within 24-30 hours. The moment larvae become free swimming, infusorians may be provided for first feeding. The fry will rich maturity within 6-8 months.

Brachydanio frankei (Leopard Danio)

A small sized fish with leopard like spots on the gold background. The dorsal area is yellowish olive. Wild *frankei* has no colour in any of the fins. This is a very active species, constantly swimming in upper strata of the aquarium. They are available in central and eastern India. This can be bred in large numbers by photoperiod method in aquarium or concrete tanks. They can be bred like *Brachydanio rerio.*

Puntius conchonius (Rosy Barb)

Wild rosy barbs are dull in colour. But the male dorsal body surface is slightly pink in colour and body is slimmer, whereas the female is dull in colour.The maximum size attained is 12 cm. Sexual maturity attained at 10th month. These fishes can be bred in large numbers in hapas with a bottom mesh of 5mm or by providing plastic filament substratum. Eggs are slightly sticky in nature but they are egg scatterers. They can be bred commercially by using hapa method with a perforation of 5mm bottom mesh. Male and female ratio can be 1: 1 or 1:2. They can be bred in infusorian water produced naturally in the breeding tank.

Colisa fasciata (Banded Gourami)

The banded gourami is sometimes also called as 'giant gourami'. The body is rich brown in colour, interrupted by pale diagonal bands throughout the body. The bands and throat of a matured male is bright in colour compared to females which shows swollen abdomen during breeding season. Overall the female is dull in colour. Thread like pelvic fins are present in both the sexes. As they are bubble nest builders, they can be bred in commercial scale or pair wise depending upon the requirements.

Colisa lalia (Dwarf Gourami)

It is a hardy fish and can breath air from the surface with the help of an accessory respiratory organ-the labyrinth. Dwarf gourami matured in 12 months of its rearing period. As they are bubble nest builders, they can be bred in mass scale or pair wise depending upon the requirements.

Parlucioma daniconius (Blackline Rasbora)

This fish has got elongated body. The body colouration is dark silver at dorsal side and faint light silver in abdomen. A blue black line runs through the length of the fish. The fins of the male have a yellowish tinge and the female body shape is slightly deeper. A parallel golden line also passes along with the black line which makes the fish more ornamental. These fishes can be bred in hapa with a bottom mesh of 5mm.

Esomus barbatus (Flying barb)

The dorsal surface of the body is greenish in colour. The maxillary barbules extend up to the tip of the pectoral fin. The fishes are quite delicate and remain in surface to column layer of the tank. They show shoaling behaviour and moves in a group. They can be bred in planted tank or in a hapa made from mosquito net.

Danio aequipinnatus (Giant Danio)

The giant danio's body is pale blue with three vertical yellow lines from the gill cover to the caudal peduncle region. Females have wider body with swollen abdomen during breeding season, where as males are slightly smaller in size with tapered abdomen. Sexual dimorphism is visible by their yellow lines turn up at the beginning of the caudal fin. They are very active fish, swims at the upper surface of the water. Breeding is carried out in mass scale in cement tanks by simulating the natural environment and photoperiod.

Danio devario (Torquoise Danio)

The body is pale blue with two vertical broken yellow lines from the gill cover to the caudal peduncle region. Females have wider body with swollen abdomen during breeding season, where as males are slightly smaller in size with tapered abdomen. During breeding season the dorsal and caudal fins of females become turn to reddish yellow in colour where as in males fins are pale yellowish. These fishes can be bred pair wise in aquarium by simulating natural environment and photoperiod.

Puntius sophore (Sophore Barb)

This barb is otherwise called as 'pool barb'. It generally breeds in ponds and pools. It reaches an adult size of 18 cm and a weight of 70g. The dorsal body surface of male is greenish silver in colour and the body is slimmer, whereas the female is oval and a red line in the lateral line of the fish. There are two black spots visible on the caudal peduncle region and at the base of dorsal fin. Sexual maturity is attainted in one year. They can be bred in large-scale provided the tank contains adequate of floating weeds.

Puntius ticto **(Two Spot Barb)**

This barb otherwise called as tic-tac-toe barb, breeds in ponds and tanks. It reaches an adult size of 10 cm and a weight of 8g. The male dorsal body surface is greenish silver in colour and body is slimmer, whereas the female is oval and a deep red line on the lateral line of the fish. The ticto barb is silver and gold with two black spots *i.e* before the pectoral fin and near the caudal peduncle region. The sexual maturity attained at one year. They can be bred in large-scale in hapas fixed in a tank with a bottom mesh of 5mm. This can also be bred by providing floating weeds.

Badis badis **(Chameleon Fish)**

They are smaller in size, reaching upto 50-70 mm. The males have a concave curve to their ventral abdominal area and the females are convex. They show a surprising amount of color changing in their tank. When they are frightened they fade to a faded color. But usually their fins are very attractive in colour. Female shows bands on their body. When the dominant male is dominating over female, he turns dark with black stripes running vertically down his sides. Since they are pit breeders, they can be provided with hiding spaces with broken PVC pipes.

Puntius fasciatus **(Melon Barb)**

Small fishes of size 50-60 cm are seen in Kabani,Vythiripuzha, Chaliyar, Kunthi, Bhavani, Kallar, Mundakkayam and Aralam wild life sanctuary located in South India. They are very colourful fishes and have much better look than ther exotic tiger barbs. Sexual maturity is attained after 18 months of rearing and average fecundity is found to be 13-15 eggs/g body weight. They can be bred in planted tanks or simulating natural environment in tanks.

Puntius filamentosus **(Filament Barb)**

They are Western Ghat species, endemic to the Southwest Indian states of Kerala, Tamil Nadu and Karnataka. It is also found in both fresh and brackish waters of rivers, estuaries, coastal marshes and reservoirs. The males are having slender body and bright colour in maturity compared to females. Female's abdomen is more round

Breeding of *Puntius filamentosus*

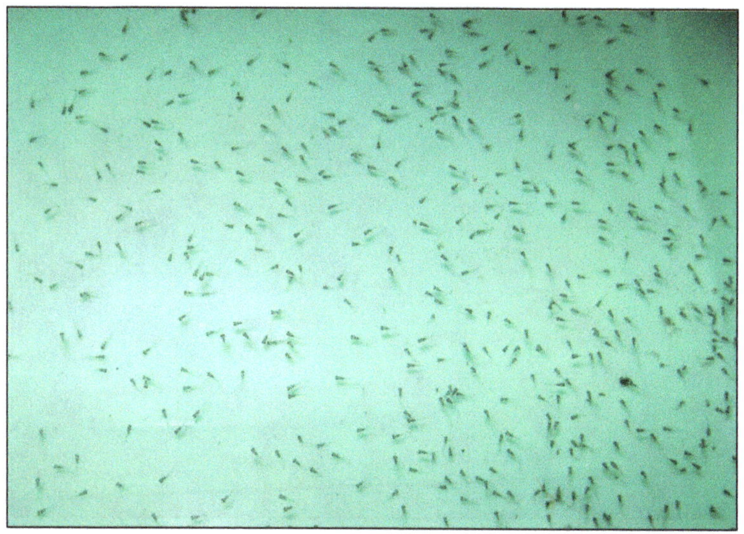

Larvae of filament barb

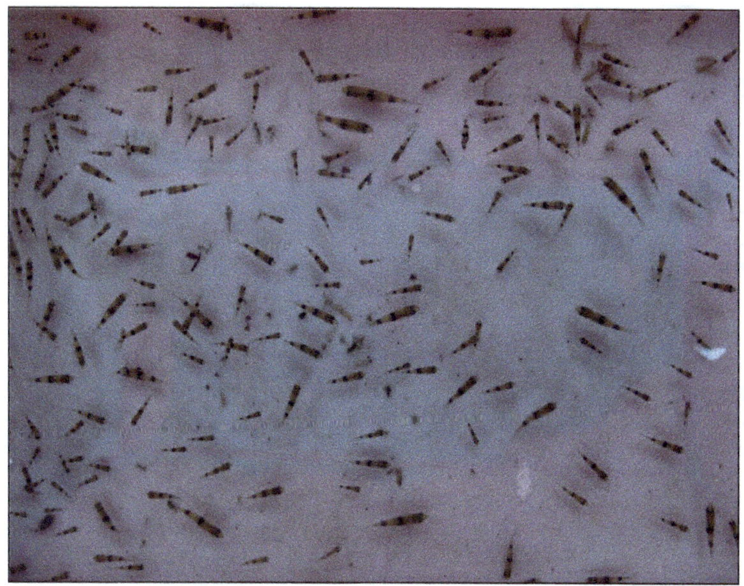

Fry of *Puntius filamentosus*

Advanced stage of Filament barb

Brood stock of *Puntius filamentosus*

Puntius filmentosus

Breeding of *Parlucioma daniconius* (*Blackline rasbora*)

Brood fish of *Blackline rasbora*

Breeding setup for *Blackline rasbora*

Parlucioma daniconius

and comparably dull in colour. Female brood fish matured under captivity with the indigenously prepared brood stock diet. The spawning was effected by the administration of hormone at a very low dose, in breeding hapas with male-female ratio of 3:2.

Breeding of *Puntius fasciatus*

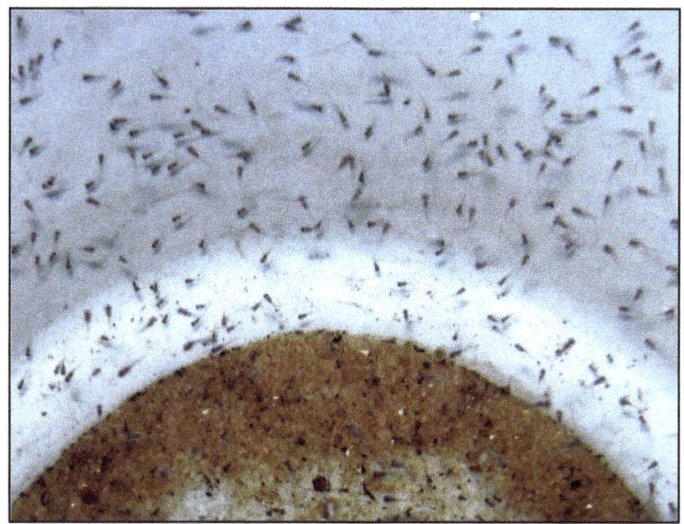

Larval stages of *Puntius faciatus*

Brood fish of *Puntius fasciatus*

Hatchery Established by CIFA in North East Region

Several programmes related to ornamental fish are taken up to popularize and motivating the local people in the states like Meghalaya and Arunachal Pradesh, Assam and Tripura state to start farming some of the important species. The large number of ornamental fish biodiversity is untapped at many places of NE

Breeding of *Puntius conchonius* (Rosy Barb)

Puntius conchonius

Breeding facilities for *Puntius conchonius*

region. CIFA regularly conducts the training programmes, awareness programmes to conserve and breed the important species. Four demonstration hatcheries specially designed for the breeding and rearing purposes are established at Killing, Meghalaya; State fish farm, Emchi, Itanagar, Arunachal Pradesh; Brood fish farm, Nagaland and Ecology and environment Department, Assam University, Silchar. Between 2000- 2009, the Institute has trained over 350 people of the NE region. To popularize the ornamental fish breeding and culture more than 1500 people have been trained from all over the country including NE region by the Ornamental fish breeding and culture unit of CIFA.

Other Ornamental Fishes

Esomus barbatus

Puntius sophore

Colisa lalia **in mass scale production**

Colisa lalia

Colisa fasciata

Badis badis

Strategies for Conservation of Wild Species

When the issue of conservation is addressed to, it is more-or-less directed towards the species that are 'typically endangered'.

Brood Fish of *Leopard danio* (*Brachydanio frankei*)

Brood Fish (*Brachydanio rerio*) Zebra Fish

However, in the context of the indigenous self-recruiting ornamental fish species, conservation of all the three categories of endangered species needs to be addressed. Suggestions for scientifically tackling these issues are stated below.

- ☆ After gathering knowledge on the habitat, ecology, feeding and reproductive behaviour of the target species, their

Danio aequipinnatus **(Giant Danio)**

Danio devario **(Torquoise Danio)**

captive breeding strategies could be scientifically thought of, methods be devised and process standardised.

- ☆ Successful rearing of larvae, young ones and juveniles in captivity.
- ☆ The food and feeding behaviour of each developmental stage of the target fish must be studied and food and feed preferences recorded.
- ☆ Training can be provided to the fish collectors for responsible utilization of the fishery resources and to also reduce the loss of fish due to poor handling while fishing, acclimatization procedures, proper handling methods and packing techniques. This would ensure eco friendliness and to improve the quality of fish.
- ☆ Eco-labeling/Green certification of the exported fish in international market is necessary. This could link up the fish traders to the international market and promote brand development of Indigenous ornamental fish.
- ☆ During rainy (breeding) season, the harvest of wild species for commercial purpose must be strictly regulated and legislated. This shall ensure a steady and sustainable supply of the species from natural environment.
- ☆ In line with the concept of social forestry, social fisheries (participatory natural resource conservation strategy) could be devised by the central Institutes in collaboration with the various central and state fisheries organisations. This would provide legal power to the local communities for the protection of the living natural resource from being subjected to excessive exploitation.
- ☆ Indiscriminate mine explosions, poisoning of natural water bodies, *viz.*, hill streams (natural flowing waters) etc. for easy harvest of food fish in the ecologically vulnerable areas are potential precursors for the extinction or at least endangerment of valuable self-recruiting indigenous ornamental fish.

Conclusion

Sustainable harvest of wild ornamental fish populations is possible, if the fish collectors and administrations have sufficient knowledge and foresight to the future disaster of overexploitation. There are certain fish dealers who only collect a limited number of specimens and fully acclimatise them before export. But, still there are others who exploit the natural resources to the fullest extent, where upon the issue of endangerment issue arises. To understand the fact that the natural resource is ours, the environment that we live in is ours, and we being the caretakers, the safeguard of nature and its resources should be our concern. The most possible solution to the issue of biodiversity conservation is to collect a limited number from the nature, feed and culture them in captivity, study their behaviour and life cycle, breed them and further utilise them for trade and other economic activities or release them safely into the wild nature.

Chapter 5

Wild Caught Native Freshwater Ornamental Fishes of India

J.D. Jameson (Late)
Retd. Director of Research and Extension (Fish)
Tamil Nadu Veterinary and Animal Sciences University,
Chennai – 600 051

Mostly wild caught fishes from the Indian freshwater bodies are exported as natural produce. Several species of the genus *Puntius* more popularly known in the aquarium hobby as *Barbus*, viz *chola, conchonius, gulio, terio, ticto, filamentosus and vittatus*, while these are predominantly plain silvery in colour and cannot hold their own against the flashy varieties like neon tetra or jewel cichlid and their advantage lies in their incessant activity and peaceful behaviour. An aquarist who keeps a couple of each species of fish has no idea how pretty at shoal of a dozen or more of these scintillating barbs can look. Even the plain black and white zebra fish (*Brachydanio rerio*, shows its real beauty in a shoal. Other Indian freshwater fishes which are well known internationally are the catfish *Mystus tengara*, the dwarf and giant gouramis, *Colisa lalia* and *Colisa fasciata*, and the Malabar danio, *Danio malabaricus*. Some of the native Indian fish species identified for aquarium hobby, recreation, culture and export are dealt with in the chapter.

Puntius arulius

It has olive green colour on back, becoming silvery white decked with reddish green over the abdomen. Three irregular black blotches are merging on body. Dorsal, caudal and anal fins appear pinkish with a black bar across the summit of the first, while the caudal fin is stained at its edges.In Tamil Nadu Nilgiri and in Kerala Wynad Kottayam ranges, and also in the Cauvery river at Srirangapatnam, Pabanasam and Manimuthar of Tirunelveli district.

Puntius chola

It has silvery, opercles shot with purple and gold, a black blotch on the tail. A dark mark is seen along the base of the anterior dorsal ray, and a row of dark spots along its center. Occasionally there is a dark mark behind the gill–opening. The fishes are distributed from Malabar and Wynad, through Chema, Orissa, Punjab, Bengal, Uttar Pradesh, Madhya Pradesh and Assam.

Puntius conchonius

It has a large, round, black spot on the middle of the side of body above the posterior portion of the anal fin. Fins are orange, dorsal with its upper half blackfish. Found in Assam, lower Bengal, Bihar, and Punjab Deccan.

Puntius denisonii

It has silvery, with a black band, above which runs a horizontal scarlet stripe passing from the snout to the center of the base of the caudal of the base of the caudal fin. Caudal fin with an oblique black band crossing the posterior third of each lobe. It is distributed in the freshwater sources of Kerala and West Bengal.

Puntius filamentosus

The adult fish has a silvery white with a deep black, oval mark on the caudal peduncle. Caudal fin red, tipped with black. Dorsal fin-rays free along boarder (young) reddish, with four black vertical bands. It is found distributed from Canara down the western coast and along the base of the Nilgiris, Manimuthar and Tambiraparani of Tirunelveli district in Tamil Nadu.

Puntius gelius

It has reddish brown, with black band over the base of the caudal fin, and another less distinct, close to the base of that fin. A silvery band is found along the side. The black peritoneum appears like an irregular dark blotch. A black spot passes across the base of the anterior half of the dorsal fin, extending one third the distance up the rays.

Puntius melanampyx

The body has a deep dull red, with three black cross bands the first from the whole of the base of the dorsal fin to just beneath the lateral line, the second commences four scales beyond the posterior extremity of the base of the dorsal fin and descends to one scale below the lateral line, while the last is just before the base of the caudal fin. Fins pinkish, edged with black. It is found distributed along Wynaad, Nilgiri and Travancore ranges and also at Cauvery. A colour variety also is found at Kodaikanal of Tamil Nadu.

Puntius narayani

It has two dark vertical bands one descending to the pectoral fin, the second across the free portion of the tail. The silvery body is suffused with golden yellow.

Puntius phutunio

The body is reddish brown with a black band passing from the back to opposite the middle of the pectoral fin, a second from the back to the posterior end of the base of the anal fin. Two other lighter bands pass downwards, one from the anterior, the other from the posterior extremity of the dorsal fin. A dark band is found down the center of the dorsal fin, another at the base of the caudal fin. They are available in Ganjam, Orissa and Bengal.

Puntius terio

The body has silvery, greenish across the back, each scale having a number of fine black spots. The fish has a large black blotch in the middle of the side over the posterior extremity of the anal fin, sometimes, extended in the median line as far as the tail. A black blotch in the young under the posterior extremity of the dorsal fin, passing downwards to the middle of the fish, becomes indistinct in the adult. Fins yellowish, their margins stained with black, with the dorsal fin having a median band. The fish is found distributed in Orissa and Bengal to Punjab.

Filament Barb
Puntius filamentosus

Tambiraparani Barb
Puntius tambraparnei

Long Snouted Barb
Puntius dorsails

Malini's Barb
Puntius Mahecola

Sophore's Barb
Puntius sophore

Kooli Barb
Puntius vittatus

Syke's Shark
Labeo boggut

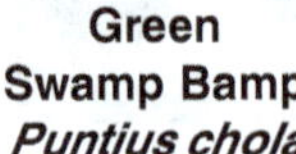

Green
Swamp Bamp
Puntius chola

Four Spot Barb
Puntius conchonius

Burjor's
Brilliance
Chela
dadiburjori

Blue Dotted Hill
Trout
Barilius bakeri

Zebra Danio
Brachydanio
rerio

Malabar Danio
Danio devario

Olivaceous Loach
Schistura devdevi

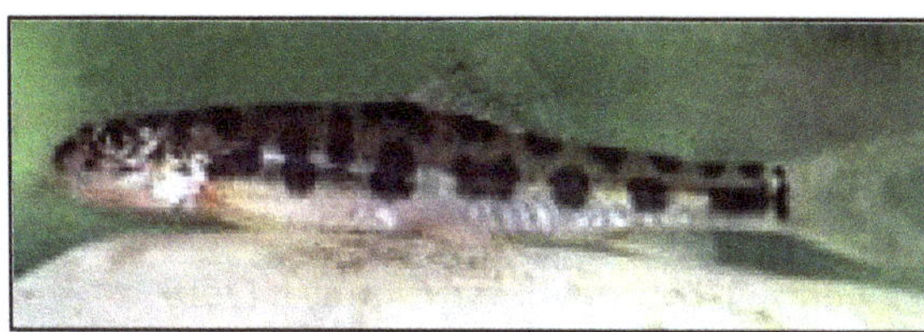

Polka Dotted Loach
Schistura croicka

Blank Line Rasbora
Parluciosoma labiosa

Puntius ticto

It has silvery sometimes stained with red, a black spot on caudal peduncle, a smaller one at the commencement of the lateral line. Fins are black and sometimes orange. They are found throughout India.

Botia almorhae

The body is reticulated with grey on a yellow background, fins yellow, Dorsal, pectoral and anal fins with four transverse black bands, the pectoral fin and each caudal lobe with five. Sometimes reticulated vertical bands are present. It is found abundantly in Kashmir, Almora and Khasi Hills.

Botia dario

Seven or eight oblique bands descend from the back to the abdomen and two or three, or even more, cross the lobes of the caudal fin.

Botia geto

The body is with irregular and partly confluent brown cross bands, which enclose variously sized round yellowish or bluish sports. Ventral fins with two, the other fins and each lobe of the caudal fin is found with three black cross bands. Found in Punjab, Himalayas, Valleys of Ganges, Jamna and in some rivers of Assam.

Namacheilus evezardi

It has greenish with small dark blotches having a vertical direction; caudal fin has four V-shaped bands. Found distributed in the river sources of Pune.

Gagata cenia

(Young) Body yellowish bronze, becoming silvery on the abdomen. Three dark bands are found over the head, four more over the back, descending as low as the lateral line. Caudal fin is with a semi lunar black band or a black blotch on each lobe. A dark mark is seen across the dorsal fin. Found in Bengal and Orissa, Jumna, Ganges and Indus rivers.

Gagata itchkeea

The body has yellowish bronze, becoming silvery on the sides and abdomen, some dark blotches along the back descending to half way down the sides. A black blotch on either lobe of the caudal fin, and another on the dorsal fin are present. These are distributed in the rivers of the Deccan.

Mystus tengara

The body is brilliant yellow, with a black shoulder spot and five black longitudinal lines. Found in Northern India, Punjab and Assam.

Somileptes gongota

It has an undulating band along the side of the body, giving off vertical bars towards the back, or else oblique blotches with light edges descending from the back, or placed irregularly on the body. Dorsal and caudal fins show transverse rows of blackish dots. These are available in the watery regions of Assam and Khasi hills.

Nangra punctata

Body is coppery glossed with gold on the sides, a black blotch on the occipital, and three or four along the back descending halfway down the sides. A black band on the dorsal fin and some black markings are found on the caudal fin. These are found available in Beerbhoom in Bengal.

Nangra viridiscens

The body is glossy greenish brown on the back, with two very light green bands, passing one from the base of each dorsal fin to the middle of the depth of the body. A dark band is found on the dorsal fin, and spots on either lobe of the caudal fin. Found in Northern Bengal, Jamna at Delhi and Pune.

Hemirhamphus leucopterus

It has a slivery lateral band, with black beak. Found in Mumbai regions.

Hemirhamphus xanthopterus

It has a brilliant lateral band two-thirds as wide as a scale in its widest part. End of beak is coral red incolour. Found in Malabar coast.

Hemirhamphus limbatus

It has a brilliant silvery lateral band, which becomes as broad as one scale. Found off the Coromandal coast.

Badis badis

Its coloration is most variable, from rose pink to earthy brown to bluish violet to grayish green, but always showing a vertical chain-link pattern. Found all through out India.

Nandus nandus

Greenish brown brassy reflections vertically marbled with three broad patches are found. A fourth one crosses the free portion of the tail.The fish is found distributed all through India.

Macropodus cupanus

It is rifle green, and has the prolonged ventral fin-ray scarlet. Caudal fin barred in sports, as is also the dorsal fin, more especially its soft portion; a round dark spot at the base of the caudal fin. Two horizontal dark bands, one from above the orbit to the upper part of the caudal fin, the other from the angle of the mouth through the eye to the lower part of the same fin, head and cheeks spotted. Found in Malabar and Coromandel and Canara.

Ctenops nobilis

The fish is golden brown, with a silvery white band, usually interrupted, passes from the eye to the middle of the tail, a second similar one from the pectoral fin along the side, and a third at the base of the anal fin. Sometimes, there appears a black, light-edged ocellus at the upper part of the base of the caudal fin. Found in North-eastern Bengal, Assam.

Colisa chuna

It is dull greenish, lighter along the abdomen; a dark, sometimes black, band along the side to the lower half of the tail. It has a dark band in the upper third of the dorsal fin, and another along the base of that fin, anal fin banded similar to dorsal, a dark band along the base of the soft portion. Caudal fin sometimes with a black at its base, its last third rather dark, occasionally with two or three transverse bands. Found from Assam to Kolkata.

Colisa lalia

It is vertically banded with scarlet and light blue, half of each scale of either colour. Dorsal and caudal fins barred in scarlet dots. Anal fin is with a dark band along its base, and a red outer edge. Found distributed in the Ganges and Jumna, Kolkata and Chennai

Colisa fasciata

Greenish above, dirty white below, a green spot on either gill cover, fourteen or more orange bands descend obliquely downwards and backwards from back to abdomen, ventral fins edged with red and variegated with black, greed and white. Dorsal and caudal fins spotted with orange. Found distributed along Coromandal coast as for South as River Krishna, Ganges, Assam and Punjab.

Ophiocephalus punctatus

Back of the fish is greenish, becoming yellow on the sides and abdomen, with a dark stripe along the side of the head. Several bands from the back pass downwards to the middle of the body. Fins are spotted. It occurs all over India.

Rhynchobolella aculeata

Brownish or greenish marbled superiorly, becoming yellowish along the abdomen, a light band along the body just above the lateral line. A series of three to nine large black ocelli, having a white or beff edge, found along the base of the soft dorsal fin. Caudal fin is with six to eight vertical brown bars, fins otherwise grayish. Pectoral fins yellow. These are available in India except Malabar.

Mastacembelus pancalus

The fish has greenish line along the back, yellowish beneath many yellowish-white spots over the sides. Posterior portion of body often vertically striped. Soft dorsal, pectoral, caudal and anal fins yellow, with numerous black spots. They are found in large rivers of India except in South of Krishna.

Mastacembelus armatus

Rich, brown, lighter on abdomen. A blackish band through the eye is continued in an undulating course along upper half of side of body. Sometimes a row of black sports occurs along the base of the soft dorsal fin, and short black bands over the back under the region of the dorsal fin spines. Pectoral fin usually spotted, dorsal and anal fins usually banded or spotted. It is found occurring all over India.

Esomus danricus

It has a board black lateral band and long maxillary barbells extend up to the anal fin. It is found distributed in India, including Nicobar Islands.

Rasbora daniconius

A black band edged with gold passes from the eye to the base of the caudal fin. Found all through out India.

Danio devario

The fish is greenish above, silvery white below. The anterior part of the body is reticulated in its centre by steel-blue lines, divided from one another by narrow vertical yellow bands. Three blue lines, divided by yellow ones, is continued backwards to the caudal fin, where the lower two amalgamate and passing upwards, become lost on the superior half of the fin. Orissa, Bengal, Deccan, Punjab and Assam.

Danio malabaricus

Back steel-blue, some irregular vertical yellow lines are found on the fore part of the body and three or four blue bands along the sides, the central ones coalescing so as to form a broad bluish band along the middle of the caudal fin. Found all along the West Coast of India.

Danio aequipinnatus

Body of the fish is yellowish white. A wide bluish band extends along the body from the eye to the centre of the base of the caudal fin; in its course are sometimes round silvery spots. Below it is another narrow band which occasionally joins the central one in the anterior. There are two other lighter bands above the central one. The intermediate ground colour is yellow. Fins are yellowish. Dorsal and anal fins each with a broad bluish band along their outer half. These are available in Himalayas at Darjeeling and whole of Assam, Naga and Garo hills, Deccan.

Danio dangila

Back olive, abdomen silvery, sides with several narrow blue lines, which in the anterior half or two-thirds of the body forms a beautiful network. A dark spot behind the gill covers. The fish has anal fin with two or three blue stripes. Bengal, Bihar and Himalayas at Darjeeling.

Brachydanio rerio

Four metallic blue lines along the sides, separated by three narrow silvery ones, and forming three bands on the caudal fin. Dorsal fin is noticed with a blue edging. Anal fin is found with three longitudinal blue bands. The fish is found distributed in Bengal and in the lower Coromandal Coast at Machhalipatnam.

Lavbuca dadiburjor

The fish has silvery with a black lateral band along the body on which, four to five black spots are seen. It is found distributed from Goa, Kochi to water bodies of Kanyakumari district.

Aplocheilus lineatus

It is greenish with a gloss of purple on the cheek and along the abdominal surface. A golden green spot is found in the centre of each scale. Eight to ten vertical black bands pass down the sides to the abdomen. Fishes are found in Coorg and Wynaad, down the Malabar Coast.

Aplocheilus panchax

Its upper surface is greenish, becoming dull white on the sides and beneath. Fins yellowish, lower third of dorsal fin covered with a large black spot. Dorsal, caudal and anal fins margined with black. It is distributed in Orissa through lower Bengal and Andaman Nicobar islands.

Barilius barna

The adult is dull green. Nine to eleven vertical, dark bands on body dorsal and caudal fins edged with black. The young have the back grey the sides silvery shot with gold, seven to nine narrow deep blue vertical bands. Fins yellow, dorsal and caudal stained with black. These are available in Assam, Ganges and its branches, Bengal and Orissa.

Barilius shacra

It is back olive, rest of the body pinkish silvery, about twelve incomplete bars from the back downward towards the lateral line, a dark bar along the upper third of the dorsal fin. Found from Haridwar down the valley of the Ganges in Assam.

Barilius bendelisis

The body is silvery, shot with purple, back of a slaty grey, descending bars towards the lateral line. In adults these bars become indistinct, but each scale with a black spot at its base, and two on each scale forming the lateral line. Fins whitish tinged with orange. A grey margin is found to the dorsal and caudal fins. Found in Assam and Himalayas through India as far as the Western Ghats, but not from Malabar and Canara.

Rohtee cotio

The body is silvery, darkest along the back and sometimes with a silvery lateral band. These are available throughout India except the Malabar Coast and South of the Kistna.

Labeo calbasu

Body blackish in colour. In clear streams many of the scales have a scarlet centre. These are available in Punjab, Kutch, Deccan, Southern India and Malabar, from the Kistna through Orissa and Bengal.

Tetrodon cutcutia

The body is greenish yellow above, white on abdomen. A light band from eye to eye. A large black ocellus surrounded by a light edge, on side. The whole back marked with dark greenish reticulations enclosing carmine. A red spot is seen on throat. These are available in Orissa, Bengal and Assam.

Tetrodon fluviatilis

It is greenish olive above, white on sides and below, back and sides with large black blotches much wider than the ground colour, sometimes quite black beneath. The fins are yellowish, and of caudal stained dark and sometimes with black spots. Found all through out India.

Syngnathus kalyanensis

It is a brownish fish of attractive value to consider as ornamental fish. Found in Kalyan in Maharashtra State.

Several freshwater Indian ornamental fish species are well known in the international arena and have been kept in European and American countries for over 70 years and many more species occur in good numbers and are being exported to these countries at present also. Native ornamental fish species of Indian waters are known for attractive coloration, peaceful native, and ready to accept artificial and live feed, adaptability to the confirmed water conditions with nominal care.

The commercial methods for the ornamental fishery are largely based on long-term partnerships among fishermen, middlemen, exporters and importers. The issues concerning improvement in the nature ornamental fishery can be addressed with direct involvement of all interested parties. Increasing local participation in the fishery and trade processes are essential to the long term sustainable use of the ornamental fishes, and to ensure the retention of revenues at the production center to raise local life standards.

The survival of wild caught ornamental fishes in India will depend not only on its sustainability, but also on the "green" value of the fishing. Amazon riverine source, the large potential greener source for ornamental fish stocks spells the phrase–'buy a fish, save a tree".

India has not made her presence felt in the International market due the reasons stated below:

- ☆ Non availability of breeding stock (being a restricted item for import)
- ☆ Lack of training in scientific breeding techniques, nutrition and healthcare.
- ☆ Inadequate fish transport facilities.
- ☆ Poor marketing strategies.
- ☆ Improper handling and packaging.
- ☆ Lack of professional training.
- ☆ High air freight for Export.

Some Indian fishes, though not unknown in the hobby, is available only to the connoisseurs, as they are not exported in large numbers. These are the stripped and "Y" loaches (*Botia striata* and *Botia lohachata*), orange and green chromides (*Etroplus maculates and E.suratenisis*), *Aplocheilus lineatus*, etc.

The advantage of a mild climate, suitable (soft) water, abundant natural food and reasonably cheap labour are encouraging the trade. Thus both wild collected as well as bred fishes can be exported from our country.

A challenge to the more enterprising exporter is the introduction of fishes which are at present not known in the international aquarium fish hobby. Already we have a few such examples, notable of the melon or emperor barb (*Puntius melanampys*) and the honey gourami (*Colisa chuna*) several species of the genus, such as *Botia, Danio, B.gulio, B.almorhae,* some dwarf catfishes of the genera *Gagota* and *Nangra*, some barbs like *Puntius sahyadrensis or P.denisonii,* etc.

Promising fish species that are of interest to the Indian exporter, together with their brief colour descriptions and location are to be explored as candidate species to enter into the trade and studying the biodiversity. The locations of marine fishes are not given, as these, especially those which are not coral fishes, extend over wide areas all over our coasts.

Ornamental fish species and their aquatic environments are overlooked in conservation and sustainable development in India. The diversity of native ornamental fishes (Over 200 Indian species) and the socio-economic value of the fisheries have great importance for the country. India represents 1 per cent of the value of ornamental fishes in the world market ($ 360 million). It is the need of the hour to conserve and maintain the live ornamental fishery at commercially feasible and ecologically sustainable level. It may also help the villagers to make their livelihood through judicious exploitation of the indigenous varieties of fish species.

Part III

Ornamental Fish Production Technologies and Management

Chapter 6

Advanced Grow-Out Systems (Raceways and Lined Ponds) for Intensive Rearing of Ornamental Fishes

S. Felix

Fisheries Research and Extension Centre
Directorate of Centre for Animal Health Studies, TANUVAS
Madhavaram Milk Colony
Chennai – 600 051

Using conventional systems such as smaller tanks, tubs, etc for ornamental fish farming is becoming outdated. Inorder to produce more and to make more profits, advanced culture systems are to be adopted in ornamental fish farming. Recirculation systems, raceway systems, zero water exchange techniques, green water technology, organic farming are few of such advanced farming practices which can be introduced in this sector to enhance fish production. This would enable the sector to grow to a different level in our country. Two of such systems designed and practiced by our University for fish/shrimp farming are detailed here.

1. LDPE Lined ponds
2. Raceway tanks

LDPE Lined ponds

Earthen ponds are provided with lining on sides and bottom with the help of low density polyethylene (LDPE) or high density polyethylene (HDPE) sheets. The advantages is that the high rate of water seepage from earthen ponds could be arrested by the lining. Further, the pumping cost also is considerably reduced in such systems, as we need to fill up water only to compensate the water loss due to evaporation.

Lining Materials

LDPE Sheet

LDPE materials are available in different thickness ranging from 150 to 300 GSM (grams per square metre). They are costing less compared to HDPE sheets and they may lost for 4-5 years, if exposed to sunlight and more than 10 years, if covered with clay/sand.

HDPE Sheet/Synthetic Fibre Spread

HDPE sheets/synthetic fibre sheets are available in different thickness ranging from 500 -1000 GSM. Relatively costlier but long lasting, would lasts for more than 10 years even if it is exposed to sunlight.

Selection of Lining Materials

The inner lining material for the pond is selected based on the following factors:

1. Soil type
2. Cost of lining material
3. Cost of installation
4. Durability
5. Suitability

Method of Sealing the Polyethylene Sheet for Lining the Pond

1. The surface of LDPE sheet to be used will be cleaned. It should be free from grease, dust and other dirt for better sealing.
2. Teflon pad will be used beneth the sealing surface for effective sealing.
3. According to the thickness of the LDPE/HDPE sheet the

heat is regulated in the sealing machine.

4. To avoid damages due to excess heat a Teflon sheet also will kept over the sheet before sealing.
5. After sealing Teflon sheet will be removed and wiped with wet cloth.
6. Finally the effectiveness of sealing has to be checked.

Raceway Ponds

Design and Construction of an Intensive Raceway System for Mass Culturing of Ornamental Fishes

Using raceways enclosed in green houses has much more advantages to that of constructional ponds and tanks. The production period can be reduced in the grow out ponds. In other words, extension of the growing season utilizing enclosed greenhouses as nurseries decreases the period required for the production of marketable-size fish in ponds. This "head starting" can result in production of more crops a year. For regions with a year-round growing season, the use of an intensive nursery management strategy will increase survival and predictability.

This chapter provides a system description with the significance of such components involved in the raceways based upon the design developed at FC&RI, Thoothukudi.

System Description

The description of nursery raceway system given in this chapter is for the unit of two raceways.

Reservoir Pond

A reservoir working volume of 100 m^3 is recommended for the two raceway unit with a total water volume of 120 m^3. To exclude large predators, reservoir intake pipes should be equipped with two filter bags of 420 and 800 ìm screens. The reservoir can be designed to feed raceway pumps by gravity to avoid constant priming problems. An external swivel standpipe at the deep end is recommended for draining and reservoir overflow control.

Greenhouse Facility

The greenhouse facility currently in use at the Fisheries College and Research Institute, Thoothukudi has solid M.S. Iron framing with high salt resistance. The roof and end walls are made of translucent corrugated fiberglass panels. The sidewalls are made of green shading (80 per cent) materials. The structure is large enough to maintain two 45^3 capacity raceways with pumps and rapid sand filters. The greenhouse space is equipped with floodlight fixtures arranged in three rows. The structure can be constructed using the locally available, durable and cost effective materials.

Layout for Raceway

Efforts are to be made to build the raceways with adequate elevation to allow fishes to be drained during harvest. Raceways can be built using concrete, fiberglass, plywood, or simply by lining a trench in the ground with high-density polyethylene (HDPE) membrane. Elongated raceways (30 m long; 3 m wide) with rounded end walls are adequate. Raceway walls can be vertical or with a 2:1 slope to the bottom. Each raceway will have a 25 m long central partition. A minimum water depth of 0.5 m and a bottom slope of 0.5 per cent with a 0.3 m deep sump at the deep end are suggested to improve waste removal. Freeboard of at least 0.15 m is needed for raceways. For raceways designed to produce larger-size fishes, deeper freeboard or net cover will be needed to avoid fish loses due to jumping.

Pump, Pipe, Lines and Valves

Each raceway should be separately provided with a pump with a capacity of about 25 m^3/h. The pump will serve to optimize oxygen injection, enhance raceway water circulation, to filter water and to pump algae from one raceway to another. The pump is used to both bring in new water and recirculate water within the raceway. The raceway pump should have access to two water sources: the reservoir and the raceways. A multiport valve (M) can direct water from the pump to the sand filter (SF), bypass it, or sent it to drain. Filling the raceways is done by the raceway supply pipe. This pipe is equipped with a 50 mm check valve and a 20 mm PVC ball valve. This valve can be used temporarily for post larval acclimation. At the shallow end raceway, the pipe has three 50 mm PVC ball valves. One valve controls water flow into the raceway, the second valve controls water flow into bottom manifold and the third valve feeds the common distribution pipe. This pipe is equipped with a bleed valve for flushing the pipe network.

Raceway water intake pipe also can be equipped with flow meter to facilitate daily water management.

Raceway Outlets and Filter Pipes

Each raceway drain outlet is positioned at the centre of the sump, half-way between the end of the partition and the raceway's end wall. The raceway water level can be controlled by a swivel external standpipe. Each raceway should be provided with a set of

Lined Ponds for Mass Production of Ornamental Fish

Figure 6.1: Laying of LDPE (250 gsm) Sheet

Figure 6.2: HDPE (700 gsm) Lined Pond

Figure 6.3: Lined Pond Using Synthetic Fibre

four filter pipes of the following screen sizes: 600, 800, 1000 and 2000 ìm. Filter pipes are mounted on the outlet to avoid losing fishes during water exchange. These are changed as the fishes grow. Two types of filter pipes are needed: the all-perforated/slotted type.

Rapid Sand Filters

Conventional swimming pools sand filter with manual backwash and filtration capacity of about 20 $m^{3/}h$ is needed for each raceway. The sand filter can be used to filter the incoming seawater as wells as the raceway water.

Multiport Valves

The multiport valve is a multiposition valve with six operational modes: sand filter, backwash, rinse, circulation, waste and closed. The backwash and rinse modes serve to maintain the sand filter in optimal working condition. The circulation bypasses the sand filter. The waste mode can be used to drain the raceway without using the external standpipe. The closed mode is a safely position to avoid accidental drainage through the multiport valve when the raceway is not in operation.

Oxygen Injection

The liquid oxygen system includes a storage tank, pressure regulator, distribution pipe and valves, oxygen filters, oxygen flow meter, and oxygen diffusers. The distribution system can deliver

Raceway System for Intensive Farming

Figure 6.4: Spreading of HDPE Sheet in Raceway Tank

Figure 6.5: Fixing of Baffle in Raceway System

Figure 6.6: Raceway Under FRP Sheet and Greenhousing

Figure 6.7: Raceway Under HDPE Sheet Housing

Figure 6.8: Raceway System(Double) in Operation

Figure 6.9: Raceway System (Single) in Operation

oxygen into an oxygen diffuser or an oxygen injection probe. The oxygen diffuser, although less efficient in terms of oxygen transfer than the probe, is not pump dependent. The probe, on the other hand, is more efficient, but requires the use of the pump. Different oxygen injection systems are available on the market, and one should choose the most appropriate method for each location.

Bottom Manifold System of Raceway Tank

The bottom manifold is a pipe network equipped with spray nozzles which is used to stir and suspend particulate mater that has settled on raceway bottoms. Each raceway is equipped with four 350 mm PVC manifold pipes, two at the centre (on each side of the partition) and one along each sidewall. All the bottom manifold pipes are interconnected for single valve operation. Manifold pipes are equipped with non-metallic, flattened-fan–shaped spray nozzles mounted in 45° PVC elbows.

Blowers and Air Distribution System

About 400 lpm of air will be needed for a nursery unit of three raceways. This air supply should be adequate for feeding a total of 50 mm airlift pumps, 1 m micropore air diffusers and 1000 litre *Artemia* hatching tanks. Air can be provided by any hig-volume, low-pressure blower. A smaller unit should be maintained as a backup. A main air distribution pipe should feed six longitudinal pipes which provide air for the raceway. Each airlift pump and air diffuser is connected with a flexible house to a valve.

Airlift Pumps

Airlift pumps are the vital component in a raceway. Water circulation in the raceway depends primarily on airlift pumps. In this device, air is introduced near the bottom of a vertical pipe which has a 90° PVC elbow at the upper section. The air, lift the water in the pipe and sends it through the elbow. Air pressure, airlift pipe diameter, submergence depth, and the type of airlift pump being used are among the main factors affecting pumping rate. Use of the collar-type, air is forced through a ring of small-diameter eight banks, four on each side of the partition, with three airlift pumps in each suspension across the raceway. This support can be built from PVC, fiberglass or wood with required weights for adequate positioning of the airlift pumps.

Air Diffusers

Each raceway is equipped with six 1 m-long air diffusers, made up of low-pressure, 25 mm rigid porous plastic pipe, providing air bubbles of 20 to 40 ìm in size. Air is provided via two rigid PVC elbow to ensure continuous airflow. Air diffusers are attached to both sides of the partition just above the bottom manifold pipe.

Lighting Facility

The greenhouse is equipped with sufficient light fixtures to provide sufficient light for night feeding without distribution shrimps and for helping with the night harvest. Lighting may also will be helpful during cloudy and rainy days to sustain algae in raceways.

Temperature Control

Building nursery facilities under greenhouses in temperate climates will help gain about 10° C above the ambient water temperature, depending the greenhouse efficiency, sunlight and water exchange. In places where the water temperature during early spring can drop occasionally below 10°C, space or water heaters are needed to maintain adequate water temperature during cold-weather periods. In tropical countries the greenhouse helps to maintain temperature in raceways. It protects the system from sunlight, dusty wind and rainfall and help to maintain the temperature.

Backup Generator

A backup generator with sufficient capacity to support the main raceway blower in case of power failure is indispensable. This generator should be equipped with an automatic switch for immediate start of blower in case of power failure.

Raceway Operation and Management

Filtered water is filled up in the raceways and with the airlift systems water is allowed to move at a regulated speed, both vertically and horizontally which enable the floating particles (feed, fecus, organic matter, microbes, plankters, etc) in suspension, and not allowing them to settle down. The mechanical agitation and the colloidal particles kept in suspension are the cordinal principal of raceway operation which help raising the fish crops successfully.

Addition of specific microbial concentrate (as probiotic) and algal concentrate (as bioremediators) takes care of the water quality and prevent the nitrogenous wastes from exceeding the optimum levels in the culture system.

Hence increased level of stocking density of fry can be adopted in such system without affecting their survival and growth of stocked fishes.

Advantages of Raceways

- ☆ Very high stocking density can be possible in such systems.
- ☆ Can be managed as 'zero water exchange system'.
- ☆ Raceways adopt the 'green water technology' helps to reduce the supplementary feed cost.
- ☆ Optimum water quality management, means less stress and free from disease outbreaks.
- ☆ For mass culturing of all kinds of ornamental fishes these systems are highly suitable.
- ☆ Through initial investments will be there, it would be economically viable.
- ☆ Proven culture systems for improving fish production–all over the world.
- ☆ Successfully designed, fabricated and adopted to Indian conditions.

Chapter 7

Aquaponics System: Integration of Aquaculture with Vegetable Hydroponics

Ravindra D. Bondre

Marine Biological Research Station

(Dr. Balasaheb Sawant Konkan Krishi Vidyapeeth)

Ratnagiri – 415 612, M.S.

As a prerequisite to genuine development, human population require high-quality protein, vitamins and minerals for proper nutrition. Fish and vegetables are excellent sources of these food groups. Therefore, there is a strong need for establishing a technology which allows for the efficient cultivation of these food groups while conserving freshwater and land resources.

Aquaponics is known as the integration of hydroponics with aquaculture. In aquaponics, nutrient rich effluent from fish rearing tanks used to fertilize hydroponics production beds and in turn, the hydroponics beds functions as a biofilter–stripping of ammonia, nitrites, nitrates and phosphorus, so the freshly cleansed water can be recirculated back into the fish rearing tanks (Diver, 2006). One of the advantages of aquaponics systems is that water replacement can be theoretically lowered, since nutrient build-ups are not high because plants uptake the nutrients.

Hydroponics is a technology for growing plants in nutrients solutions (water and fertilizers), with or without the use of an artificial medium (*e.g.* sand, gravel, vermiculite, rock wool, peat moss, saw dust) to provide mechanical support. Advantages of hydroponics are (*i*) density resets planting, maximum crop yield, (*ii*) crop production where no soil exists, (*iii*) more efficient use of water and fertilizers and (*iv*) minimal use of land area and disease control (Jenson, 1991).

Background

In aquaponics system, water is replaced (not exchanged) only to account for the water lost through plant-mediated evapo-transpiration. Researchers showed that waste nutrients could be stripped off from fish culture waters using hydroponically grown plants, with the hydroponic component generally using a sand gravel culture beds.

A number of investigations have been carried out into the use of hydroponics for maintaining water quality in closed recirculatory fish production systems. It has been reported that by using gravel bed hydroponics showed that good water quality was maintained and in gravel bed hydroponics tomatoes who yields higher than the plants grown in soil. However, supplementary nutrients were supplied to provide the desired nutrients balance.

A report shows the feasibility of using a small nutrient film technique (NFT) in which hydroponic unit served as a mineral filter for a recirculating aquarium system. A recirculating NFT system was designed using a 40-L capacity aquarium, and stocked with common carp, *Cyprinus carpio* and rainbow trout, *Onchorynchus mykiss*. Four plant crop species were grown (garden peas, *Pisum sativum*, lettuce, *Lactuca sativa*, watercress, *Nasturium officinate* and barley, *Hordeum vulgare*) with no additional nutrient supply. He found that if the hydroponic unit is incorporated into the aquarium filter system as a single passage unit, water quality could be maintained in optimal units, together with the production of a useful crop.

The choice of hydroponics growing system within an aquaponics context may be based on the independent advantages conferred by that particular hydroponics component. For example, sand/gravel systems may remove the requirement for a separate bio-filter, as the substrate will also act as a substrate for nitrifying bacteria, and therefore replaces conventional bio-filters. Similarly the gravel sand substrates may also act as a solid filtering medium.

A reciprocating flooding and draining cycle for the hydroponics bed was thought to be essential to the efficient operation of the system as reciprocation provides uniform distribution of nutrient rich water within the media during the flood phase and improved aeration in the media during the drain component of the cycle due to effective atmospheric exchange of gases in the drained hydroponic bed. Both the nitrifying bacteria and plant roots are expected to benefit in this system. From the earlier in their comparative studies of reciprocating flow versus constant flow in an integrated-gravel bed-aquaponics system showed that constant flow treatment appear to remove more nitrate from culture waters than did reciprocating control treatments, but no significant differences was detected. Similarly, final phosphate levels within reciprocating control treatments were higher than that in constant flow treatments, but no significant differences was detected.

A comparison has been maded between three different hydroponics systems (gravel bed, floating and nutrient film technique) in an aquaponics system, with Murray Cod; *Maccullochella peeli peeli* (Mitchel) in fish culture tank and green oak lettuce; *Lactuca sativa*, a leafy vegetable plant in hydroponics subsystems. Lettuce yields were good, and in terms of biomass gain and yield, followed the relationship gravel bed > floating > NFT with significant differences in all treatments. The NFT treatment was significantly less efficient than the other two treatments in terms of nitrate removal (20 per cent less efficient).

Advantages of Integration of Aquaculture with Vegetable-Hydroponics

- ☆ The aquaponics system is extremely conservative of the indigenous natural resources, especially of freshwater supplies, it is deemed most appropriate for: application in the region with inadequacy supply of freshwater; areas with soils unsuitable for the practice of traditional aquaculture and agriculture methodologies.

- Vegetable hydroponics are incorporated/integrated with fish culture to recover nutrients that would otherwise accumulate or be discharged into the environment.
- Vegetables are valuable additional products of fish production that enhance system profit potential.
- Although nutrients are not major expenses in commercial hydroponics, most essential plant nutrients derived from feed accumulate in fish culture at no additional expense.
- Integrating fish culture and vegetable hydroponics can offer additional savings through shared infrastructure such as pumps, heaters, reservoirs, monitoring and control, and vertical space in greenhouses.
- Nutrient removal by plants improves the quality of fish culture water and may enhance fish production.
- Plant roots, hydroponics structure and media improve water quality by capturing solids and providing surface area for bio-filtration.
- The purpose of implementing this integrating system had several fold:
    1. To demonstrate and establish a comprehensive, environmentally - friendly food production system.
    2. To establish the sustained capacity to maximize food production efficiency per unit input of valuable natural resources.
    3. Reduce seasonal fluctuations in availability of vegetable and fishery dietary products.

Components of Integrated Aquaculture and Vegetable Hydroponics

Although integration of aquaculture with vegetable hydroponics (referred as 'aquaponics systems' hereafter) configurations is varying greatly, an optimum arrangement of unit process components has emerged as Figure 7.1. Effluents from the fish-rearing tank are first treated to reduce suspended solids and biological oxygen demand (BOD). Next, culture water is treated to oxidize ammonia in a bio-filter by nitrification. Then water flows through the vegetable production unit where plant uptake recovers dissolved nutrients. Finally, water is collected in a reservoir from

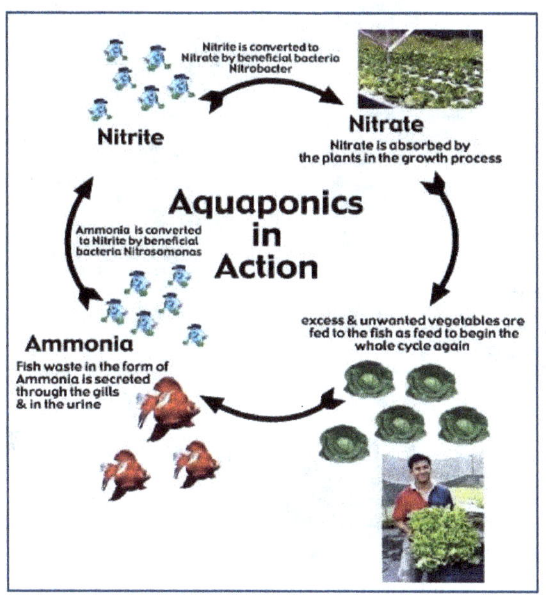

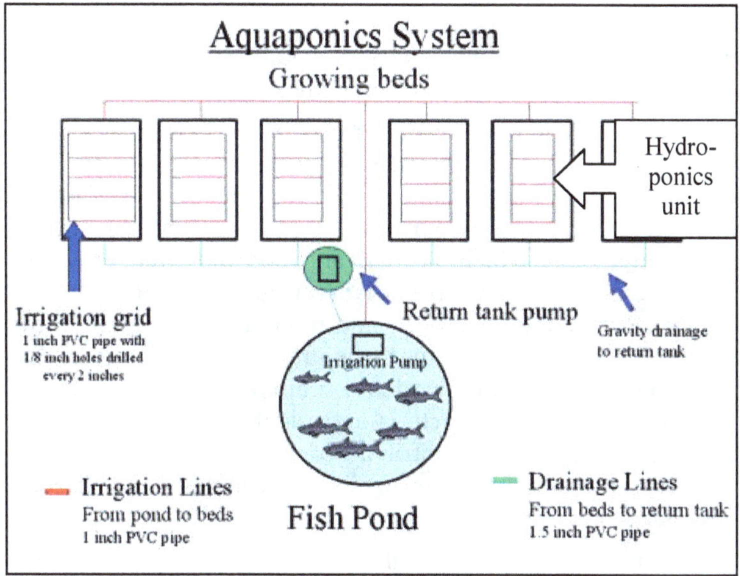

Figure 7.1: Aquaponics: Principle and a Model Farm

Figure 7.2: Demonstration of Aquaponics Technology at Faculty of Fisheries, Maharashtra

which it is returned to the fish-rearing tank or in case of vertical system, water is returned gravitationally to the fish-rearing tank.

In the integrated system, once the hydroponics area (biofilter) is flooded during a pumping cycle, the withdrawal of waste-laden aquaculture water ceases and the hydroponics area (biofilter) is allowed to drain the cleansed water back into the 'tank'. Hydroponics area, which alternatively flood and drain, provide i. uniform distribution of nutrient-laden water within the filtration medium during the flood cycle and ii. improved aeration of the hydroponics area (biofilter) due to the atmosphere exchange created by each dewatering. This 'flood-and-drain' water movement cycle benefits both nitrifying bacteria and plant roots. The increased availability of oxygen to the nitrifying bacteria improves their ability to convert ammonical-nitrogen to nitrate (facilitates the nitrification process).

Scientists compared three different hydroponics systems (gravel bed, floating and nutrient film technique) in an aquaponics system, with Murray Cod; *Maccullochella peeli peeli* (Mitchel) in fish culture

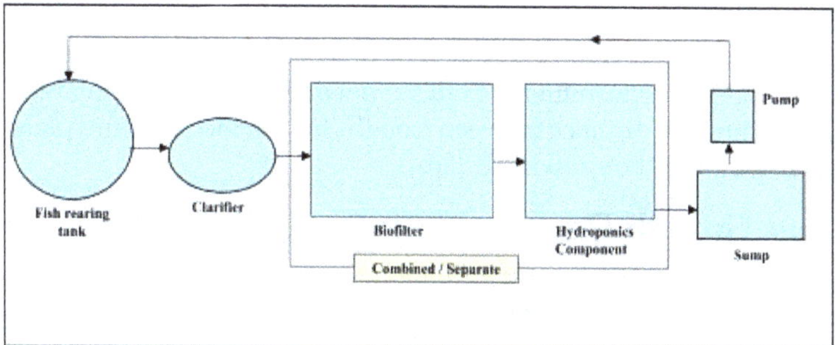

Figure 7.3: Components of Integrated Aquaculture and Vegetable Hydroponics

tank and green oak lettuce; *Lactuca sativa*, a leafy vegetable plant in hydroponics subsystems. Lettuce yields were good, and in terms of biomass gain and yield, followed the relationship gravel bed > floating > NFT with significant differences in all treatments. The NFT treatment was significantly less efficient than the other two treatments in terms of nitrate removal (20 per cent less efficient).

Aquaponics System: Component Ratio

Aquaponics systems are generally designed to meet the minimum size requirements for the solids removal and bio-filtration components as they relate to fish production. The purpose of designing aquaponics systems should be the high efficiency of water use in production of quality food as well as high functional and technological simplicity. Practical considerations (*e.g.* space, technical demand on management and economics) are also important factors in the establishment of the configuration and relative size of aquaponics systems' components.

Different Hydroponic Systems Used in Aquaponics System

Of the wide range of hydroponics systems used in integrated culture, inert substrate systems have been frequently used with gravel culture being the most common. To ensure adequate aeration of plant roots, gravel beds have been operated in a reciprocating mode, in which culture water is applied continuously to the base of individual plants through small diameter plastic tubing. Nutrient film technique (NFT) has been successfully incorporated into aquaponics systems.

NFT consists of many narrow plastic troughs in which plant roots are exposed to a thin, flowing film of nutrient solution applied continuously or intermittently. High plant density can be maintained by adjusting the distance between troughs to provide optimum plant spacing during the growing cycle.

Plant Growth Requirements in Hydroponics

Maximum plant growth in aquaponics systems requires proper nutrition consisting of six macronutrients (N, P, K, Ca, Mg and S) and seven micronutrients (Cl, Fe, Mn, Zn, Cu, Mo and B). High dissolved oxygen concentration in the water surrounding the plant root is critical for maximum nutrient uptake and growth. Excessive solids or stagnant water in the root zone leading to oxygen depletion will cause root damage. Effective solids removal and aeration of the root zone are major design and operational considerations of the hydroponics systems.

It is reported that a broad range of vegetable species have been shown to assimilate sufficient quantities of all required elements derived solely from the fish waste products, and there have been no nutrient toxicities seen. Yield responses are generally favourable over a relatively wide range of fish culture tank: hydroponics area (biofilter) ratios but vary by species. Fish yield rates tend to be relatively suppressed at low plant population densities per unit of fish cultured.

Other important aspect for hydroponics vegetable production is climatic factors. Production is generally best in regions with maximum intensity and duration of sunlight.

Water Quality Management

Dissolved oxygen concentration of > 5 mg/L in the fish-rearing tank are required for maximum growth. After culture, water is discharged from the fish-rearing tank, suspended solids are reduced. Sedimentation is frequently used to remove settleable solids (>100 m) from aquaponics systems. Following reduction of suspended solids and BOD, culture water passes to a biofilter designed primarily for the oxidation of ammonia and nitrite by a fixed film nitrification process. Biofilters for nitrification perform optimally within a temperature range of 25 - 30° C, a pH range of 7.5 - 8.0 at a saturated dissolved oxygen concentration, low BOD concentration (< 20 mg/L) and total alkalinity of 100 mg/L as $CaCO_3$ or greater. A number of investigators have documented the contributions of vegetable hydroponics to water quality improvements through the determination of nitrogen and phosphorus recovery. Alkalinity declines in fish - vegetable co-culture systems as bicarbonate is consumed and acid is formed during nitrification. Liming agents are added to maintain pH at 7.5 for integrated plant - fish culture.

Fish and Vegetable Species

Several species of fish, ranging from warm water to coldwater species have been cultured in aquaponics systems. Warm water temperature (25 - 30°C) will generally promote high vegetable growth and production, thus integration with culture of warm water fish species is more appropriate.

A wide range of vegetables has been grown in aquaponics systems. Most studies have assessed the productive potentials of high value cash crops such as tomatoes, lettuce and cucumbers. All of these crops proved to be suitable for aquaponics systems. Leafy vegetables, like lettuce, are particularly suitable for aquaponics systems. A crop can be produced in a short period (6 - 8 weeks) and as consequences, disease and pest infestation is relatively low.

Needs and Prospects of Aquaponics Systems

Research on aquaponics systems during last 15 years has been exploratory in nature, involving considerable trial and error. Nevertheless, quantitative relationships are starting to emerge and there is a clearer understanding of principles, practical approaches, and promising designs. The outlook for increasingly productive development of integrated system is encouraging. The immediate potentials for integrated system products appears to be niche markets where consumers are willing to a higher price for consistently high-quality fish and produce available on a regular basis throughout the year. Through environmental control integrated system operations will be able to locate adjacent to urban markets, substantially reducing transportation costs. Aquaponics systems also have potential in arid and semi arid areas due to water reuse efficiency.

Aquaponics systems are found to be designed and managed to meet the production requirements of both fish and vegetables. To create the optimum interface, the normal range of design and management options must be narrowed considerably. Chemicals that are approved for application to vegetables cannot be used with fish nor can approved fish therapeutants be used with vegetables. Production is to be restricted to those species and varieties that can either resist pest and diseases or to be protected by biological control methods. Finally, the development of integrated system faces financial constraints in the form of the cost of investment capital for commercial implementation as unproven and high-risk ventures always have difficulty in obtaining investment capital.

Economic aspects of aquaponics systems have received little attention. Economic evaluations are ultimately necessary to secure investment capital, thereby facilitating the application of information generated from laboratory based research to commercial development of an integrated system. An increase in scale of experimental system will more accurately simulate realistic operating conditions.

With considerations of aquaponics systems that have potentials, an attention is required in a number of critical areas. Development of the optimum interface between fish and vegetable system requires further development of engineering criteria for system design and operation. The nature of the relationships between daily feed input, system volume and plant production area for different vegetables requires elaboration. The optimum relationship will maximize yield and nutrient recovery and minimum water exchange and nutrient accumulation, thereby extending the duration of water reuse.

Conclusion

In India, in view of increasing population, it is now imperative to utilize the available land and water resources judiciously. In this context, aquaponics–integration of hydroponics with aquaculture, can one of viable alternative for maximum sustainable fish and plant production together. The major advantages of aquaponics system is limited consumption of water; the system could not only be a large industrial plant but also easily be adopted at small scale such as home aquarium (ornamental or food fish) combined with a mini garden, for growing herbs, vegetables or flowers.

References

Bohl, M., 1977. Some initial aquaculture experiments in recirculating water systems. *Aquaculture*, 11: 323–328.

Bender, J., 1984. An integrated system of aquaculture, vegetable production and solar heating in an urban environment. *Aquacultural Engineering*, 3: 141–152.

Clarkson, R. and Lane, S.D., 1991. Use of small-scale nutrient film hydroponic technique to reduce mineral accumulation in aquarium water. *Aquacult. Fish. Manage.*, 22(1): 37–45.

Diver, S., 2006. *Aquaponics: Integration of Hydroponics with Aquaculture.* A publication of ATTRA–National Sustainable Agriculture Information Service. Available on the web at www.attra.ncat.org/attra–pub/PDF/aquaponic.pdf

Jenson, M.H., 1991. *Hydroponics Culture for the Tropics: Oppurtunities and Alternatives.* Department of Plant Sciences, University of Arizona, Tucson, Arizona, U. S. A. (Available online at http://www.agnet.org/library/abstract/eb329.html)

Lennard, W.A., Leonard, B.V., 2004. A comparison of reciprocal flow verses constant flow in an integrated, gravel bed, aquaponic test system. *Aquacult. Int.*, 12: 539–553.

Lewis, W.M., Yopp, J.H., Schramm, H.L. and Brandenburg, A.M., 1978. Use of hydroponics to maintain quality of recirculated water in a fish culture system. *Trans. Am. Fish. Soc.*, 107(1): 92–99.

Seawright, D.E., Stickney, R.R. and Walker, R.B., 1998. Nutrient dynamics in integrated aquaculture-hydroponics systems. *Aquaculture*, 160: 215–237.

Watten, B.J. and Busch, R.L., 1984. Tropical production of Tilapia (*Sarotherodon aurea*) and tomatoes (*Lycopersicon esculentum*) in a small scale recirculating water system. *Aquaculture*, 41(3): 271–283.

For more information and extensive bibliography, please follow the link: http://attra.ncat.org/attra–pub/aquaponic.pdf

Chapter 8

Ornamental Arowana

V. Sundararaj[1] *and D. Yuvaraj*[2]

[1]*Consultant: Aquaculture, Environment and Biodiversity Chennai – 600 073*

[2]*Asian Analytical Laboratories Pvt Ltd, Chennai – 600 041*

Ornamental fishes are regarded as a consumer based commodity and they are star products of the pet markets. Ornamental fisheries is a multi-billion dollar industry that supports thousands of rural people in the developing countries. The global ornamental fish trade in retail level is worth more than US$ 8 billion with an average annual growth of 9 per cent, while the entire industry including plants, accessories, aquarium, feed, medications, etc. is estimated to be worth around US$ 20 billion.

Earlier studies made in the 1980's showed that gazing at colourful aquarium fish reduces stress and subsequently lower blood pressure. Due to the stresses of urban life, more and more people are seeking therapeutic, calming activities as hobbies. The growing awareness of the need to respect the environment and to know more about it, has also helped popularize recreational activities with an educational component. Keeping fish in an aquarium caters to all these needs and it has become popular among people from all walks of life.

India's Favourable Conditions

In a tropical country like India, where environmental conditions are more favourable, ornamental fish culture could be a very valuable enterprise both in the freshwater and marine situations for production and uplift of rural areas. Ornamental fish species of these two extremely different ecosystems are good in their biodiversity and selected species can be commercially farmed adopting modern techniques and marketed.

While considering the price of freshwater ornamental fishes, it is about 10 times higher than edible fishes. The costs of marine ornamental fishes are believed to be 10 times greater than freshwater ornamental fishes. Thus, by value, normally the attraction should be more towards the marine ornamental fishes. But the reality is not so due to practical reasons. Maintenance and breeding of freshwater ornamental fishes is easier than marine ornamental fishes. It is worth mentioning here that 95 per cent of the marine ornamental fishes in the globe are collected from natural environment and only 5 per cent is due to breeding and production.

Ornamental Fishes: Their Features of Attraction

Ornamental fishes have certain specific features of attraction and because of them hobbyists are more attracted, concerned and prepared to pay any amount to get their preferred fish. Usually, the attraction of the ornamental fishes is due to their body shapes, dignified movements, bright colours, fancy fins, kind of scales, gentle movement, graceful look, calm behaviour, living together, love making, kissing mouth, etc.

Among the spectrum of ornamental organisms, fin fishes are the dominant ones and are commanding over the invertebrates due to the following merits.

1. Beautiful colour
2. Gentle and active movement
3. Accepting artificial feeds
4. Peaceful nature
5. Different body forms

The major (97–98 per cent) organisms in the aquarium are the fin fishes in terms of varieties or species diversity. Among the ornamental fishes, 77 per cent are freshwater fishes and 20 per cent are marine fishes. The contribution of others is only 2-3 per cent.

Arowanas

Arowanas are a special group of bony fishes, living in freshwater with certain adaptive characteristics, especially for aerial respiration. Arowanas come under the class Actinopterygii, order-Osteoglossiformes and family Osteoglossidae. This family has 2 sub-families, namely Heterotidinae (having two genera) and Osteoglossinae (having two genera). Thus, arowanas have four genera (*Arapaima, Heterostis, Osteoglossum* and *Scleropages*). The first two genera have one species each. The third genus, *Osteoglossum* has two species and the last genus, *Scleropages* has six species. Thus the species diversity of arowanas as could be seen by the classified are ten.

Family Osteoglossidae

Subfamily Heterotidinae

- ☆ Genus *Arapaima*
- ☆ Arapaima or pirarucu, *Arapaima gigas.*
- ☆ Genus *Heterotis*
- ☆ African arowana, *Heterotis niloticus.*

Subfamily Osteoglossinae

- ☆ Genus *Osteoglossum.*
- ☆ Silver arowana, *Osteoglossum bicirrhosum.*
- ☆ Black arowana, *Osteoglossum ferreirai.*
- ☆ Genus *Scleropages*
- ☆ Red-tailed golden arowana *Scleropages aureus.*
- ☆ Green arowana or gold crossback arowana, *Scleropages formosus.*
- ☆ Gulf saratoga, red saratoga or northern spotted barramundi, *Scleropages jardinii.*
- ☆ Red arowana, super red arowana, or chili red arowana, *Scleropages legendrei.*
- ☆ Saratoga, silver saratoga or spotted barramundi, *Scleropages leichardti.*
- ☆ Silver Asian arowana, *Scleropages macrocephalus.*

Ornamental Arowana – Desirable Species

Scleropages formosus

Scleropages legendrei

Scleropages aureus

Scleropages leichardtii

Scleropages macrocephalus

Arowanas, are known as aruanas or arawanas. In this family of fishes, the head is bony and the elongate body is covered by large and heavy scales, with a mosaic pattern of canals. The dorsal and the anal fins have soft rays and are long based, while the pectoral and ventral fins are comparatively small. The name "bonytongues" for arowanas is derived from a toothed bone on the floor of the mouth, the "tongue", equipped with teeth that bite against teeth on the roof of the mouth. The fish can obtain oxygen from the atmospheric air by sucking it into the swim bladder, which is lined with capillaries like lung tissue. The arapaima is an "obligatory air breather".

Distribution

Among the ten described living species of Arowana, three are from South America, one is from Africa, four are from Asia, and the remaining two from Australia. Thus the species are widely distributed, but restricting to freshwater bodies.

Behavior

Arowanas are carnivorous, often being specialized surface feeders. They are energetic and excellent jumpers. *Osteoglossum* species have been seen, leaping more than 6 feet from the water surface to pick off insects and birds from overhanging branches in South America. Arowanas have been rumored to capture prey as large as low flying bats and small birds. All species of arowana are large, and the Arapaima is one of the world's largest freshwater fish, attaining 2.5 metres in length and weighing around 200 kg.

In the Aquarium

Usually, arowana is kept as a single fish in an aquarium. Since arowanas tend to merge in groups of five to eight, they can also be kept as group in sufficiently big tanks. Any fewer numbers may show an excess of dominance and aggression. Some compatible fish, which have often been partnered with this fish are listed below.

Costly Arowana

Though marine ornamental fishes, particularly the most colourful ones from the coral reefs are costlier, certain freshwater ornamental fishes are dictating their heavy price. They are costlier. For example, a fingerling of arowana is sold at the rate of Rs.2000–4000 per individual and the adults are commonly sold from Rs.30,000 to 40,000 for several reasons, especially based on the faith on them that fortune will come and health will be protected if arowana is kept in aquarium at home. Further, species of arowana are really graceful to look at. In certain trade fairs, specially conducted for arowana, the selected and awarded ones have been auctioned and the maximum rate has been upto Rs.7,00,000. This would indicate the love of the people on arowana.

Hardy Arowana

Arowana being a hardy fish, can be easily maintained at homes, offices, Institutes and public places. Further, they can also be had in garden ponds and interested hobbyists can reproduce them and earn. Above all, Asian arowanas can be successfully cultured in farm ponds also. Anybody venturing in the farming activity will be able to generate good income by sale/export of such fish.

Farming Countries

Arowana farming activities are going on in countries like Taiwan, China, Malaysia and Singapore. Concerned Governments encourage the interested persons to proceed firmly for farming. A student of Malaysia went to Taiwan to study about ornamental fishes, especially arowana. After successful completion of his professional studies, he started a farm of his own and demonstrated his potential. Now his farm is the biggest for arowana production. When other tropical countries are involved in arowana farming, India too can do it!

Arowanas show territorial habit and that must be taken into account when introducing tank mates. Details of the possible tank mates are provided below.

Tank Mates for Arowana

1. Sting rays: *Potamotrygon motoro, P. humerosa, P. reticulate*
2. Gars: *Lepisosteus osseus, L. ocellatus, L. leopoldi*
3. Bichirs: *Polypterus senegalus, P. ornatipinnis*
4. Others: *Distichodus sexfasciatus, Flower horn, Asttronotus ocellatus, Channa* sp., *Osphronemus gourami, Mastacembelus* sp.

Arowana–Biological Significance

Arowana (sometimes called dragon fish) can be a great choice to those who think big. Some Arowana varieties can grow upto four feet long (120cm). They can be feisty, though become tamer with age to the point of eating from your fingers, and not the fingers themselves. Some varieties are nicknamed "Bony Tongued Fish".

Arowanas are carnivores, though they generally eat nearly anything. Young Arowana fish may be fed frozen or live brine shrimp, black worms and even small fish. Baby Arowana should be fed 3 times a day, medium sized twice a day, and adults once a day, or even once every other day. Variety food is vital for a well balanced diet to Arowana fish, just like for most other fish.

Arowana, due to the eating habits, produce a lot of nitrogenous waste and therefore, one should pay extra attention to its water conditions (especially Ammonia, Nitrite, and Nitrate) in the aquarium. Changing 25 to 30 per cent of the water weekly is advisable. Generally, a good and healthy Arowana will grow upto at least 60-75cm. Some varieties can become 120 cm in the wild.

Arowana will often swim in the surface water of the aquarium and are capable of jumping from the aquarium. Hence, it is desirable to keep the aquarium well covered to avoid death of a pet fish. The Silver Arowana in the wild has been known to jump at insects in trees. Arowana may live for many years, and if well cared for, Arowana fish can live, longer than 20 years in-captivity.

Endangered

The Asian Arowana or Golden Arowana (*Scleropages formosus*) is considered an endangered species. Care should be taken to follow the law in purchasing and transporting them. Asian Arowana generally can grow to about 36 inches, and are often much more expensive than the other Arowana species. These are well known and popular in South East Asia, where they are believed to bring luck. Feeding them healthy guppies, gold fish, frogs, or shrimp makes a good stable diet. The temperature is best when kept between 24-30 °C. A pH level between 6.5–7.5 is advisable.

Arowana-Selection Criteria

Selection is a must for perfection and better performance. Ornamental arowana must also be subjected to careful selection, which must be based on, swimming posture, body shape, tail/fins, barbels, eyes, mouth/lips, scales and gill cover.

Nutritional Requirements

Fishes, as others, satisfy their food requirement mainly through natural selection. They require wide range of nutrients like protein, fat, carbohydrate (major nutrients), vitamins and minerals (minor nutrients) to perform the functions of the body. Though protein provides energy, its primary function is to provide amino acids for the body building process. Fat particularly provides essential fatty acids, while carbohydrates function as an energy source. Vitamins and minerals do not supply energy, but they play an important role in the regulation of the metabolic activity. Minerals are used for the formation of body skeleton.

Thus various components of the pointed out sources must be present in the prepared food in the required proportion, to serve as 'balanced food'. The merit of prepared food is that it can be balanced while live foods need not be so. Preference for live foods, prepared foods or both depends on the food and feeding habits of the specific species of fish.

Arowana is a carnivorous fish. It hunts for its food/pray. In fact, it is a predatory fish, with suitably modified mouth structure. In wild, its food consist mainly insects, fishes, prawns, worms and small amphibians. Live food is preferred by arowana. However, they can be trained to accept other prepared food. Variety in food is a must for arowana to have a balanced diet, and to avoid nutritional deficiency. The desired live foods of Arowana are listed below.

Live Foods for Arowana

1. Crickets (*Acheta domesticus*)
2. Cockroach (*Periplaneta americana*)
3. Centipedes
4. Guppy (*Poecilia reticulata*)
5. Fish flesh
6. Freshwater prawn
7. Marine shrimp
8. Krill (Euphasia)
9. Mealworm (*Tenebrio molitor*)
10. Bloodworm (*Chironomous larvae*)
11. Tubifex worms (*Tubifex tubifex*)
12. Earth worms
13. Frogs (*Rana hexadactyla*)
14. Tadpoles

Live food is generally more nutritious than its counterpart, the prepared/artificial food. However, the risk of introducing disease into the tank is more. This is especially true, when the live food is waterborne. Live foods should normally be quarantined for at least a few days before feeding to the arowana. Care must be taken to see that fat is not deposited around arowana's eyes due to more feeding with gold fish (which are fatty).

Water Quality and its Management

Among the various parameters, which affect fish directly or indirectly, there are major ones, affecting the fish directly. They are water quality, food and pathogens. For better fish health, maintaining water quality at low levels of carbon dioxide, nitrite and ammonia and proper pH, alkalinity and hardness are very important.

Water quality deterioration, if noticed in a tank or in farm, should be attended immediately. Increasing aeration, reducing feed rates, controlling phytoplankton bloom and water flow management are the important ones to be undertaken without any delay for abating low dissolved oxygen content. Water exchange during periods of poor water quality also helps to regain the health status of the fish in the tank or farm.

Water quality must be always maintained at optimal level stably for good health of the ornamental fishes, including Arowana. Though arowana is a hardy fish, it should be provided optimal conditions.

Optimum Water Quality Parameters

Characteristics	*Range*
O_2 (mg/l)	> 3
pH	6.5–7.5
Ammonia (un-ionised) (mg/l)	0–0.02
Nitrate (mg/l)	0–3.0
Nitrite (mg/l)	0–3.0
Phosphorous (mg/l)	0.01–3.0
Zinc (mg/l)	0–0.05
Total Hardness ($CaCO_3$) (mg/l)	10–200
Calcium (mg/l)	10–160
CO_2 (mg/l)	0–15
H_2S (mg/l)	0–0.002
Iron (total) (mg/l)	0–0.5
Manganese (mg/l)	0- 0.01
Total alkalinity ($CaCO_3$) (mg/l)	10–400
Nitrogen (gas saturation) (mg/l)	<100 per cent
Total Solids (mg/l)	50 - 500

Uneaten food, faecal waste and dead plant matter would increase the organic load in the habitat or living place.

Among ornamental fishes, Arowana is specific on certain capabilities. With regard to tolerance to water quality parameters, they are hardy. Yet, requirements of some parameters are to be met with as per the table provided.

Ornamental arowanas can be cultured in garden ponds, brood stocks developed and bred successfully. The breeding biology of arowana is interesting and the parental care, shown by the 'male' is amazing. Before culturing Arowana, one need know its biology. When one involve himself in arowana or any other ornamental fish breeding, one will realize the possibility of enjoying happiness, in these days of stress.

Chapter 9

Aquarium Fish Health Management, Biosecurity and Quarantine

K.M. Shankar
Department of Aquaculture,
College of Fisheries,
Mangalore

Health management is an important component of aquarium industry. As the aquarium is an artificial ecosystem which is subjected to a wide range of human interferences, disease of several types are expected and experienced. Diseases and health management as in aquaculture, revolves around three important components, the host, pathogen and the environment. Broadly diseases in aquaria are classified into two categories; Non Infectious and Infectious.

Non Infectious Diseases

These diseases are due to mainly alteration in environmental conditions. These are not spreading type. Altered environmental parameters stress the fish and hence have profound impact on the immune system, pathogen and spread of diseases.

Depletion of Dissolved Oxygen (DO)

Fish, feed waste, microbe and plants during night consume DO. Any change in the input particularly density/crowding and waste accumulation cause DO depletion.

Clinical Signs: Surfacing, gulping with mouth wide open

Remedy: Aeration, waste removal and filtration.

Excess Oxygen

Cause a bubble trap in blood vessels, on gills and body leading to discomfort and deaths

Remedy: Reduce plants and light

Acid or Alkaline Water

pH less than 5 and more than 9 are not desirable. pH depends on the source of water.

Clinical signs: Milky discoloration of skin due to acidosis. Ragged condition of fins and irritation to gills due to high pH.

Remedy: Neutralise acid water or change the water source. To reduce pH, reduce the plants as more plants produce more CO_2 and bicarbonate.

Temperature

Ideal temperature for warm water fish is 20-28°C. Low temperature stress the fish resulting in reduced immunity and increased infection.

Accumulation of Nitrogenous Waste

Accumulation of organic matter and fish excreta are source of ammonia, nitrite and nitrate. Ammonia and nitrite are toxic and not desirable even at sub lethal level. Accumulation of nitrogenous waste is a common in new tanks.

Signs: gill and skin damage, pale color and prone to disease.

Remedy: Establishing good bio-filtetrs, and plants for using nitrate.

Chlorine

Chlorine source is municipal water or residues of disinfection in aquaria.

Signs: Pale gills with white edge.

Remedy: Tap water has chlorine from 0.2 to.5 ppm. Store water for 48 h for dechlorination, or pass the water through activated charcoal or use Sodium thiosulphate.

Poisons

Lead, Zinc, Copper can enter aquarium through material used in pipe line and other material in tank construction.

Nutritional Imbalance

Poor health due to incorrect nutrition is a common. A balance of artificial and natural feed is necessary. Fattening of fish is common problem due to luxury feeding and less activity of fish in aquaria. Egg binding due to inability of aquarium fish to resorb egg as in normal condition is common leading to health hazard. Intestinal inflammation is another common ailment due to imbalanced feeding.

Infectious Diseases

Parasitic Diseases

External Parasites

Ich Disease

It is common disease in freshwater, brackish water and marine aquaria caused by *Ichthyopthirius* a protozoan ciliate. It is a skin parasite.

Signs: severe attack look like sprinkled grain on body, fins leading to white spots.

Remedy: Change over method to break the parasite life cycle. Swarmers can survive for 50 h outside host, Changing 12 h for a new tank. Quarantine for 1 week, Acriflavin 1 g/100 l, quinine 1g/100 l for 9-10 days, Methylene blue 1 g/100 l

Velvet Disease

Also called "Blue slime disease" is caused by Costia sp, a protozoan disease, feeding on skin and gill

Signs: folded fins, weak, rubbing body against sides and bottom of the tank. Remedy: Acriflavin 1 g/100 l for 2-3 days.

Fin Rot

Caused by several agents such as bacteria, protozoa (velvet disease), fungi (Saprolegnia) and hereditary fin rot in platys and guppies.

Signs: Fin margin damage, necrosis, inflammation, peeling of fins in bits. Remedy : antibiotics if bacterial, Aureomycin 250-500 mg/5 litre.

Odinium Infection

An algal infection of skin and gill in fresh and marine fish. It can be serious in marine fish.

Signs: Appear similar to ich disease. Very fast multiplication of the alga and spreading.

Remedy: Acriflavinn as for Ich treatment, reducing light to discourage the algae. $CuSO_4$ at 0.4 to 0.8 mg/l.

Gill Parasites

Gyrodactylus 4 sp. common on gills.

Remedy : 10 g common salt/litre for 20 min

Carp Lice

Argulus is a skin parasite with toxin gland.

Signs: Retracted fin, scratching movement.

Remedy: Lindane (Trichlorophon)

Ergasilus

Freshwater parasite on gills.

Signs: pale gill, emaciation.

Remedy: Lindane

Lernaea

Lernaea infection on body surface cause much damage and leads to secondary infection.

Internal Parasites

Plistophora Infection

It is a sporozoan a common parasite of tetras. It is a muscular parasite in deep muscle. spread through water, fish to fish and through eggs.

Signs: Paleness of colour specially in neon tetras, white patch on tail and back.

Remedy: Destruction of fish is the best option

Worms

Infection with worms such as round worms and acanthocephalan is common in aquarium fish. Affect general well being of fish.

Bacterial Diseases

Dropsy

Caused by *Aeromonas hydrophila* and Pseudomonas.- opportunistic pathogens.

Signs: scale protrusion, accumulation of ascites body fluid. popped eye.

Remedy: Antibiotics

Fin Rot

It could also be due to bacterial pathogens.

Signs: Margin of fin affected to begin with followed by inflammation and loss of whole fin.

Remedy: Antibiotics

Fungal Diseases

Saprolegnia and Achlya-wound Parasites

Infection of skin of weak and young fish. Eggs are also affected in hatchery. Secondary pathogens active upon physical injury under bad water conditions.

Remedy: $KMnO_4$ 1 g/l, Malachite green 0.2 g/l for short duration

Ichtyophonus (Ichthyosporidia)

It is one of the common and serious infection in aquarium fish. It is estimated that nearly 50-60 per cent of all deaths in aquaria are due to this fungus. It can multiply in large numbers virtually in all tissues and organs of fish. It is a stealthy dangerous disease. It spreads very fast in an aquarium.

Signs: A number of common clinical signs found with other infection such as loss of fin, skin lesion, colour change, emaciation, gasping. It can be easily diagnosed by globular blackish fungal hypae inside organs under microscope

Dermocystidium

It is a disease affecting koi. The disease mainly affects skin and occasionally the gills. It is believed to be a fungal disease although, some tend to describe it as protozoan infection. It causes raised swelling varying in size from 2 to 10 cm. The lesions are pinkish to red and vary in shape from circular to elongated ovals. There is minimal inflammatory reaction around lesions. Presence of hyphae and spores can be viewed in wet mounts, histological sections and stained smears. When lesion matures it raptures spreading thousands of spores into the water.

Viral Diseases

Koi Herpes Virus

The recent outbreaks of Koi herpes virus in the neighboring South - East Asian countries is a cause of worry to the country. Koi herpes virus (KHV) is a highly contagious virus, capable of causing significant morbidity and mortality in common carp, *Cyprinus carpio* and Koi. The virus may cause 80-100 per cent mortality in affected populations and affects fish of various ages. The virus is believed to remain in the infected fish for life, thus exposed or recovered fish should be considered as potential carriers of the virus. As large numbers of ornamental fishes, especially koi are being imported to India and there is a high risk of KHV entering in the country.

Zoonosis

It is spread of pathogen to human being through aquarium fish. Fish pathogens/organisms mostly bacteria and protozoa in aquaria have potential for zoonosis. It may be through ingestion and contact of water. Most of these pathogens are opportunists infecting immunocompromised aquarium handlers.

1. Fish Nycobacterium: A mycobacterium causing local granulomatous nodule on the skin of hand, fingers. Infection takes 6-8 weeks and can spread to lymph nodes. Aquarium handlers should use gloves, wash hands after work and explain to physicians about the type of work
2. Infection with *Aeromonas hydrophila,* and *Vibrio* sp." Ubiquitous, secondary pathogens of fish and have the potential to cause disease in aquarium handlers.

Diagnosis of Diseases in Aquarium Fishes

Diseases are detected by clinical signs and case history followed by external examination- visual and microscopic. Microbilogical examination by isolation and identification of organism also give important clue about pathogen. Molecular biological tools using

DNA and antibody probes are commonly employed today. Rapid and simple field diagnostic tools are also becoming popular.

Type of Treatments

Fish are treated by oral (through feed), injection and bath (continuous bath and short dip).

Disinfection of Aquaria

Important to break the pathogen cycle. Discord all plants, pebbles, etc. Use $KMnO_4$ 1g/10 lit for 5 days or chlorine/sodium hypochlorite.

Biosecurity

Biosecurity is a term used in animal farming industry to describe the preventive measures taken against any infectious disease outbreak. It is sum of all procedures in place to protect organisms from contracting, carrying or spreading disease. Thus, biosecurity in aquaculture is the protection of fish or shellfish from infectious (viral, bacterial, fungal, or parasitic) agents. The key elements of biosecurity are reliable source of stocks, adequate detection and diagnostic methods for excludable diseases, disinfection and pathogen eradication methods, best management practices, and practical and acceptable legislation. The program should be tailored to the needs of the specific site, business needs of the operation, the fish species and life stages grown and the disease profile of the surrounding region. Overall, a biosecurity program would include, but not be limited to, practices and procedures involving: (*i*) Surveillance for the presence of disease organisms, (*ii*) Vaccination, (*iii*) Quarantine and restricted access, (*iv*) Appropriate practices of fish husbandry, (*v*) Disinfection and (*vi*) Disease treatment (including eradication).

General security precautions need to be established in each facility to help support the activities of both disease prevention and disease control. A manual of standard operating procedures (SOP) should be assembled to provide a set of standard rules for biosecurity measures and disease monitoring. This should include such things as facility design, facility flow for both personnel and stock, rules for limited or restricted access to facility, required visitor log book, disinfection procedures for personnel and equipment, a waste management plan, pest control guidelines, and general husbandry and management procedures. Thus biosecurity is a team effort, a shared responsibility, and an on-going process to be followed at all times. From the breeder to the hatchery, to grow out operators, biosecurity measures and good aquaculture practices have to be observed to contribute to the success of the industry.

Health management, in the context of transboundary movement, is necessary to reduce the risks arising from the potential entry, establishment or spread of pathogens and the diseases. Therefore, it should encompass all activities related to the preparation, transportation and receiving of aquatic animals that are moved between regions, countries or territories. All countries reserve the right to take sanitary and phytosanitary (SPS) measures necessary for the protection of human, animal, or plant life. The appropriate level of protection (ALOP) required is determined based on relevant economic, social and ecological factors of the particular country. Countries should encourage industries to use preventative measures to limit their exposure to pathogens and disease. Such measures include the use of better management practices (BMPs), health certification, specific pathogen free (SPF) and high health (HH) stocks, quarantine, and vaccination protocols. The national aquatic animal health strategies and health management procedures to be formulated or maintained by each country should adhere to international and regional standards and be harmonized as wide a basis as possible.

Legislation

Several procedures and guidelines developed by different agencies, organizations or nations deal with the components of biosecurity issues and plans. The common objectives include aspects of protecting animal populations, environment, food and the humans itself. Many instruments falling under the terms such as policies, codes, agreements, plans, conventions, regulations and treaties have been made to achieve the objectives of biosecurity. Examples are given in the Table.

International or multinational policy instruments containing elements pertinent to aquaculture biosecurity. Dates are years of initial adoption (from Scarfe, 2003)

Lead Organization	*Title*
World Trade organization (WTO)	Agreement on the application of Sanitary and Phytosanitary Measures (SPS Agreement), 1995 Convention on Biological Diversity (CBD), 1992, and its Cartegena Protocol on Biosafety, 2000
Food and Agricultural Organization of the United Nations (FAO)	Organization of the United Nations (FAO) Codex Alimentarius (Codes of Hygienic Practice for the Products of Aquaculture), 1981-1999 Code of Conduct for Responsible Fisheries, 1995 Code of Conduct for the Import and Release of Exotic Biological Control Agents, 1995 International Plant Protection Convention (IPPC), 1997 International Council for the Explorations of the Sea (ICES) Code of Practice on Introduction and Transfer of Marine Organisms, 1994
International Maritime Organizations (IMO)	Guidelines for Control and Management of Ships' ballast Water to Minimize the Transfer of Harmful Organisms and Pathogens, 1997
United Nations (UN)	Biological Weapons and Toxins Convention, 1972
International Union for the Conservation of Nature	Guide to Designing Legal and Institutional Frameworks on Alien Invasive Species, 1999

Quarantine

Quarantine is defined as the maintaining of a group of aquatic animals in isolation with no direct or indirect contact with other aquatic animals, in order to undergo observation for a specified length of time and, if appropriate, testing and treatment, including proper treatment of the effluent waters. Quarantine programs for aquatic organisms typically involve protocols for inspection of animals for disease agents and certification *i.e.*, the issuing of a stating that a particular lot of animals or a production facility has been inspected and is free from infection by a particular pathogen(s).

Quarantine programs may be effective at several levels. At the international and national levels, these programs form an integral part of strategies aimed to protect the natural environment and native fauna from the deleterious impacts of exotic species. Within the country, quarantining helps to reduce the spread of pathogens, while at the local level, *i.e.*, at hatchery and farm level, quarantine of fry and broodstock from an outside source helps to protect the hatchery/

farm potentially devastating losses caused by disease. Quarantine facilities need to be built up at exporter's as well as importer's premises. Facilities also to be built up at all ports of entry.

The stringency of quarantine applied should be commensurate with the estimated level of risk. Quarantine procedures, including observation for clinical sings of disease and diagnostic testing, can be conducted in the country of origin, in a country of transit and/or in the receiving country.

Risk Analysis

It is an integral part of health management system. The first movement (introduction) of a new exotic aquatic species into an area often poses an unknown and potentially high level of pathogen risk, and thus such requests should be subjected to ecological, genetic and pathogen risk analyses. Such introductions will require special stakeholder consultations, including all countries sharing transboundary waters, to evaluate scientific evidence regarding the risk of introducing pathogens to new area. In cases where insufficient knowledge exists in relation to disease risks posed by a particular movement of an aquatic animal, a precautionary approach should be adopted by the receiving country.

Health Certification

It is a document declaring the health status of aquatic animals in a batch/consignment, from a country, zone or aquaculture establishment. It is issued by a competent authority of the exporting country for international trade in live or dead aquatic animal products. For internal movements of aquatic animals, the head of a disease diagnosis service, head of government lab can issue the required certificate. The formats and diagnostic tests as specified by the OIE's *Aquatic animal health code and Manual of diagnostic tests for aquatic animals* may be used for examining the animals and issuing health certificates.

Quarantine Actions

Inspection

1. Inspection of documents is performed to determine the presence of required documents, such as import permit and fish health certificates. If the required documents are present, the consignment is subjected to health inspection.

2. Inspection of consignment is performed to detect the presence of quarantinable disease. It may be done on board or after it has been unloaded from the conveyance.

Detection

After inspection if it becomes evident that required documents have been fully complied with, carriers of pest and diseases may be detained for observation at a fish quarantine establishment.

Isolation and Observation

For further detection of certain quarantinable pests and diseases, which due to their nature, require a longer period for manifestation.

Treatment

Many diseases, especially those caused by external parasites, can be treated. However, because chemical therapy can cause additional health complications, such as the development of antibiotic- resistant strains of bacteria, it should be used responsibly, with due caution and expert advice. Should a serious untreatable disease or pathogen be encountered in aquatic animals held in quarantine, the entire stock should be destroyed and the facility appropriately disinfected.

Approval Status

Introduction from sources that have passed a quarantine containment process may receive "approval" status if conditions do not change at the export site, further reducing quarantine requirements/duration.

Disease Surveillance, Monitoring and Reporting

In health management, the most current information on the diseases is extremely important. Hence, a systematic approach to gather and interpret disease information within the country needs to be developed. For the purpose, we need to establish a networking mechanism to facilitate collection of disease information from all the available sources. Disease surveillance and reporting forms the basis for all national disease control programs. It helps to meet trade requirements and forms the basis for risk analysis.

Surveillance includes both passive and active or targeted surveillance. One of the first steps in surveillance is to establish a national disease database and to use the database to aid regional and international reporting, develop disease control strategies, and identify and prioritize areas for research and capacity building. The National Bureau of Fish Genetic Research (NBFGR), Lucknow has initiated steps towards establishing a database for India. The FAO guidelines suggest that countries should provide accurate and timely reporting of disease notifications and associated epidemiological information on a quarterly basis to the World Organization for Animal Health (OIE) or to other disease reporting systems (*e.g.* the NACA/FAO, *Quarterly aquatic animal disease reports QAAD* (*Asia and Pacific Region*). To enable this, necessary diagnostics and reporting procedures should be established. Standardized field and laboratory methodologies and resource material should be developed and field personnel should be trained in disease recognition and reporting, to ensure accurate and rapid pathogen identification.

Zoning

Zoning can be a highly effective means to restrict the spread of important pathogens of aquatic animals which also aid in their eradication. Where a serious disease is present in part of a nation's territory and eradication is not feasible, countries should consider the possibility of zoning as a means to establish and maintain zones free of the disease and to permit international and domestic trade in live aquatic animals originating from these zones.

Emergency Preparedness and Contingency Planning

A contingency management plan has to be prepared well in advance so that at everyone in the framework knows their responsibilities and actions to be taken in the face of an emergency. Such emergency planning for aquatic animal diseases should form a core function of government services. In order to respond rapidly and effectively to contain and eradicate serious disease outbreaks caused by transboundary aquatic animal diseases (TAADs) and thus minimize their social and economic impacts, countries should develop and test national contingency plans.

Part IV

Marine Ornamental Fisheries: Biodiversity, Production and Management

Chapter 10

Sustainable Marine Ornamental Fish Trade: An Indian Perspective

G. Gopakumar
Mandapam Regional Centre of
Central Marine Fisheries Research Institute,
Mandapam Camp – 623 520, Tamil Nadu

Introduction

In recent years, the international trade of marine ornamentals has been expanding and has grown into a multimillion dollar enterprise. The marine ornamentals include fishes, stony corals, soft corals, sea fans, ornamental shrimps, sebellids, giant clams, ornamental echinoderms and live rocks. The ornamental animals are the highest valued product that can be harvested from a coral reef. According to Larkin and Degnar (2001) the global marine ornamental trade is estimated at US$ 200-330 million . The ornamental trade is operated throughout the tropics. Philippines, Indonesia, Solomon Islands, Sri Lanka, Australia, Fiji, Maldives and Palau supplied more than 98 per cent of the total number of marine ornamental fish exported the in recent years. India is endowed with vast resource potential of marine ornamentals distributed in the coral seas and rocky coasts with patchy coral formations. In the context of the expanding global scenario, it appears very much lucrative for India to venture into this industry. But it is a multi-stakeholder industry ranging from specimen collectors, culturists, wholesalers, transhippers, retailers, hobbyists to researchers, government resource managers and conservators and hence involves a series of issues to be addressed and policies to be formulated for developing and expanding a sustainable trade. In this context, a critical assessment of the current global scenario of marine ornamental trade can provide much insight into the multivarious issues associated with trade, which will be of much relevance while formulating policies for the development of a long term sustainable marine ornamental industry in India.

Global Scenario

Based on the Global Marine Aquarium Database (GMAD) the annual global trade is between 20 million and 24 million numbers for marine ornamental fish, 11-12 million numbers for corals and 9-10 million for other ornamental invertebrates. A total of 1471 species of fish are traded globally. Most of these species are associated with coral reefs although a relatively high number of species are associated with other habitats such as sea grass beds, mangroves and mud flats. According to the data provided by exporters, the Philippines, Indonesia, the Solomon Islands, Sri Lanka, Australia, Fiji, the Maldives and Palau, together supplied more than 98 per cent of the total number of fish exported.GMAD trade records from importers for the years 1997-2002 showed that the United States, the United Kingdom, the Netherlands, France and Germany were the most important countries of destination, comprising 99 per cent of all imports of marine ornamental fish. Exporters data revealed Taiwan, Japan and Hong Kong to be important importing areas.

Fishes

Among the most commonly traded families of fish, Pomacentridae dominate accounting for 43 per cent of all fish traded. They are followed by species belonging to Pomacanthidae (8 per cent), Acanthuridae (8 per cent), Labridae (6 per cent), Gobiidae (5 per cent), Chaetodontidae (4 per cent), Callionymidae (3 per cent), Microdesmidae (2 per cent), Serranidae (2 per cent) and Blennidae (2 per cent). For the years 1997-2002, the blue green damselfish (*Chromis viridis*), the clown anemone fish (*Amphiprion ocellaris*), the whitetail Dascyllus (*Dascyllus aruanus*), the sapphire devil (*Chrysiptera cyanea*) and the three spot damsel (*Dascyllus trimaculatus*) are the most commonly traded species.The top ten species together account for 36 per cent of all fish traded from 1997 to 2002.

During 1997-2002, *Amphiprion ocellaris, Chromis viridis, Labroides dimidiatus, Chrysiptera cyanea, Paracanthus hepatus, Pseudanthias squamipinnis* are the most commonly imported species into the EU. Together the top ten species make up 37 per cent of all fish imported into the EU between 1997 and 2002. Similarly for the United States, the top ten species including *Dascyllus aruanus, Chrysiptera cyanea, Dascyllus trimaculatus* and *Labroides dimidiatus* accounted for 39 per cent of all the species exported to the US.

Indian Scenario

India is endowed with vast resource potential of marine ornamentals distributed in the coral seas and rocky coasts with patchy coral formations. The major oceanic reef areas of coral reef distribution in India are the Lakshadweep Islands and the Andaman–Nicobar groups of Islands. The other areas of coral fish distribution are the coastal areas of fringing or patch reefs of Gulf of Kutch to Mumbai, areas of central west coast between Mumbai to Goa, certain locations of south west coast (Thirumullavaram, Vizhinjam to Kanyakumari), Visakhapatnam, Gulf of Mannar and Palk Bay. A total of 836 numbers of reef associated fishes are reported from Indian waters (www.fishbase.org).

Issues Associated with Wild Collection

Destructive collection practices, the introduction of alien species, overexploitation, the lack of scientific information on many species collected and threat of extinction of target species are the major problems of the marine ornamental fish trade. Destructive fishing techniques include the use of sodium cyanide and other chemicals to stun and catch fish. Eventhough cyanide only stuns the fishes, high post capture mortality is recorded. It may destroy the coral reef habitat by poisoning and killing non-target animals, including corals. During collection of coral pieces for the coral trade, many more colonies may be damaged or broken than are actually harvested. Corals are also broken for easy access to capture fish. This is more common with branching species in which small species such as *Dascyllus* and *Chromis* often refuge. Collection of live rock has been considered as potentially destructive as it may lead to increased erosion and loss of important fisheries habitat.

When exploitation is at a lower level in comparison to the resource available, there will not be any negative impact on reef fish populations. A study of the Cook Islands showed that the total catch per unit effort remained constant between 1990 and 1994 (Bertram,1996). In Australia, due to the permit system, the current aquarium fishery is at sustainable level (Queensland Fisheries Management Authority,1999). But Australia is a rare case as the Great Barrier Reef is the largest reef system in the world. It is well known that not all fish are equally available or equally attractive to the industry, and the most common fish need not be those favoured by the hobbyists. As a result, the effect of collection of ornamentals should be measured with respect to their potential to deplete particular species or locations. Several countries in Asia and South America have begun to implement restrictions for collections of certain ornamental fish species. Although no marine species collected for the aquarium trade have been driven to global extinction, studies carried out in Sri Lanka, Kenya, the Philippines, Indonesia, Hawaii and Australia have reported localized depletion of a number of target aquarium species of fish like butterflyfish and angelfish due to heavy collection pressure. The only systematic study assessing the effects of harvesting fish for the aquarium trade on resource populations was carried out in Hawaii. The study reported that eight of the ten species most targeted by collectors showed decline in abundance at exploited sites relative to control sites.

A larger part of the trade of ornamentals is centred on individual species. The vulnerability of the species to collection will depend on a number of life history parameters like growth, reproduction and recruitment. Eventhough reef fish exhibit a wide variety of mating strategies the larvae are distributed through wave and wind driven ocean currents. This makes replenishment of reefs with new fish larvae highly dependent on these currents and hence the availability of fish for sustainable aquarium collection is highly variable.

The effect of fishing are significantly different for species that are hermaphroditic compared with species that do not change sex. A fishery selectively removing larger animals first will mean that animals will have to start changing sex in smaller sizes, possibly reducing the fitness of individuals and thus making hermaphroditic stocks more vulnerable to overfishing.

Trade in ornamental marine fishes is characterised by extreme selective harvesting. For many species, juveniles are preferred by aquarium fish collectors due to their distinctive colouration and ease of maintenance. Consistent harvesting of juveniles may leave only limited number of young ones to reach adult size and replenish the adult stock.

Eventhough most coral reef fishes have broad distribution, a few species are endemics. Some species are naturally rare, occurring only in very restricted locations or naturally occur in lower numbers, eventhough they may be widely distributed. Other species may be abundant at different sites, but their distribution is limited to specific habitats. The more wide spread/or abundant a species is, the less vulnerable it is to exploitation. Increased rarity often implies higher prices and hence vulnerable to overexploitation.

Males of many coral reef fishes tend to be preferred due to their distinctive colouration. Selective harvesting for males of particular populations on a regular basis may lead to reproductive failure and ultimate population collapse due to heavily biased sex ratios in remaining population.

There are many factors that lead to post harvesting mortality, such as physical damage and use of chemicals during collection, poor handling practice and disease. Even when collected in an environmentally sound manner, aquarium organisms often suffer from poor handling and transport practices resulting in stress and poor health of fishes. Research on the marine ornamental trade between Sri Lanka and the UK demonstrated that in the mid 1980s about 15 per cent of fish died during and immediately after collection, another 10 per cent died during transit and a further 5 per cent in holding facilities. As a result of such mortality, more fishes are required to be collected than would be necessary to meet the market demand.

Development of a Sustainable Trade

A critical analysis of current global trade of the marine ornamentals from wild collections reveals many ecological concerns which require policy interventions. The major aspect that should receive top most priority is for taking appropriate action to ensure that the development of the trade should not threaten the sustainability of the coral reef ecosystem. The following measures are suggested.

Introduction of Certification for Wild Collected Species

Marine Aquarium Council (MAC) has developed a certification scheme that will track an animal from collector to hobbyist. Established in 1996, the goals of MAC are to develop standards for quality products and sustainable practices and a system to certify compliance with these standards, and create consumer demand for certified products. With a net work of 2600 stakeholders in more than 60 countries, it is recognized as the lead organization for developing and co-ordinating efforts to ensure that the international trade in ornamental marine organisms is sustainable. MAC certification covers both practices and products.

Industry operators can be certified through an evaluation for compliance with the appropriate MAC standard for the certification of practices. For certification of products MAC certified marine ornamentals must be harvested from a certified collection area and pass from are certified operations to another. MAC–certified marine organisms bear the "MAC-certified" label on the tanks and boxes in which they are kept and shipped.

Development of Hatchery Technologies for Selected Species

The ultimate answer to a long term sustainable trade of marine ornamentals can be achieved only through the development of culture technologies. It is well accepted as an environmentally sound way to increase the supply of marine ornamentals by reducing the pressure on wild population and producing juvenile and market sized fish of wide variety of fish year round. In addition hatchery produced fish are hardier and fair better in captivity and survive longer. Eventhough techniques are available for culture of corals, according to CITES data only 0.3 per cent of the total global trade in live coral is from mariculture. Most branching corals can be easily propagated from small trimmings clipped from a parent colony and in about a year a five to ten fold increase in biomass can be obtained. Soft coral fragments can grow to marketable size within 4–12 months and stony corals like *Acropora* within 4–6 months. More than 75 species corals are bred under captivity, but fast growing corals appeared to be economically profitable.

The list of marine ornamental fishes reared in captivity today contains more than 100 species. The maximum number of species reared are from the family *Pomacentridae.* Attempts for spawning and rearing in closed systems have proved technically challenging for most species except Pomacentrids like *Amphiprion* spp. and the existing mariculture projects have been developed on a relatively small scale. The great obstacle to successful tank breeding of ornamental reef fish is rearing larvae beyond the 6th to 8th day of development, a time typically associated with failure to initiate larval feeding.

Artificial seed production techniques are available for giant clams and hence giant clam culture has increased considerably. Now there are successful giant calm hatcheries for aquarium trade, in most tropical pacific nations and island groups. The culture of ornamental invertebrates other than giant clams and cleaner shrimps is constrained due to lack of information's on key life history characteristics.

Indian Context

In India, till date no organized trade of marine ornamentals has been initiated. But it is a fact that a great deal of illegal collection of marine ornamentals is in vogue in many parts of our reef ecosystem. No data regarding the same is available and this is a matter of great concern due to the indiscriminate nature of exploitation and ecohostile methods of collection which damage the reef ecosystem. In addition to this, lack of knowledge on appropriate post harvest husbandry practices leads to large scale mortality of the collected animals. It is time to evolve a marine ornamental fisheries policy in the country for developing an organized trade of marine ornamentals.

A critical analysis of current global trade of the marine ornamentals from wild collections reveals many ecological concerns which require policy interventions. The major aspect that should receive top most priority is for taking appropriate action to ensure that the development of the trade should not threaten the sustainability of the coral reef ecosystem. The destructive collection practices such as use of cyanide should be banned by legislation and enforced. Results from a recent study demonstrated that colonies of commonly traded species of corals and soft corals to varying concentrations of cyanide over different periods of time caused mortality in all corals. *Acropora,* the genus which is specifically targeted by fishers for collection of fish as they tend to hide amongst its braches is most vulnerable to cyanide exposure, showing rapid signs of stress and bleaching. Another aspect of concern is the impact of exploitation on population due to selective harvesting of species which are of high demand in the trade. Here also policy intervention through legislation has to play a key role. Several countries in Asia and South America have begun to implement collection restrictions on certain ornamental fish species. Although no marine species collected for the aquarium trade have been driven to global extinction, studies carried out in Sri Lanka, Kenya, the Philippines, Indonesia, Hawaii and Australia have reported localized depletion of a number of targeted aquarium species due to heavy collection pressure. Studies have also shown that removal of larger quantities of cleaner wrasses and cleaner shrimps which play key roles in reef health creates negative impacts on reef diversity. The third aspect of concern is the exploitation of species which are not suited for aquarium. This also needs to be avoided by legislation. The fourth aspect which demands regulations is regarding the post harvest mortality. Research on marine ornamental trade between Sri Lanka and the United Kingdom demonstrated that in mid 1980's about 50 per cent fish died during and immediately after collection another 10 per cent during transport and 5 per cent in holding facilities. As a result of such mortally more fish often need to be collected for meeting the market demand. Where organisms are collected, stored and handled by adequately trained individuals and transported in suitable containers fish mortality have been very low. The post harvest conditioning facilities should include modern gadgets such as UV lighting system, protein skimmers and carbon filters.

In the light of the above it is evident that while developing a marine ornamental industry in India it is inevitable to formulate legislations on these issues which are of vital concern to the sustainability of the trade. It is suggested that a few number of entrepreneurs can be licensed to collect suitable species from selected areas and trained on ecofriendly collection methods and conditioning and maintaining of harvested species in healthy condition. The Central Marine Fisheries Research Institute (CMFRI) and the National Bureau of Fish Genetic Resources (NBFGR) can combine to develop a certification system on line with standards developed by the Marine Aquarium Council (MAC). The Marine Products Exports Development Authority (MPEDA) can take the lead to develop an export market for the certified varieties. The impact of exploitation has to be closely monitored by scientific agencies at periodic intervals and required management measures have to be implemented as and when required.

Status of Hatchery Production Technologies Developed in India

It is well accepted that the trade developed from tank reared fish and other ornamentals is the final solution for a long term sustainable trade. During the past few years, the Central Marine Fisheries Research Institute and Fisheries Division of Central Agriculture Research Institute (ICAR) have intensified research activities on breeding and culture of marine ornamental fishes. One of the recent achievements is the success in the hatchery production of clown fish and few damsel fishes.

The Central Marine Fisheries Research Institute (CMFRI) has been focusing on this vital aspect for the past few years. The Institute was able to develop hatchery production methods of the following twelve species of ornamental fishes (clown fishes and damselfishes) which are in high demand in the international trade.

1. *Amphiprion percula*–Orange clown
2. *A. ocellaris*–False clown
3. *Premnas biaculeatus*–Maroon clown (spine cheek anemonefish)
4. *Amphiprion sebae*–Sebae Clown
5. *Pomacentrus cearuleus*–Blue damsel

6. *Pomacentrus pavo*–Peacock damsel
7. *Dascyllus trimaculatus*–Three spot damsel
8. *Dascyllus aruanus*–Humbug damsel
9. *Chromis viridis*–Bluegreen damsel
10. *Neopomacentrus nemurus*–Yellowtail damsel
11. *Neopomacentrus filamentosus*–Filamentous tail damsel
12. *Chrysipteracyanea* sp.–Sapphiredevil damsel

These methodologies developed can be scaled up for commercial level production and a hatchery produced marine ornamental fish trade could be developed in the country.

Research and development in the breeding and culture of marine ornamentals is a priority area which has to be intensified in the coming years. The high unit value of ornamentals makes them more commercially viable than marine food fish culture.

Conclusion

Eventhough hatchery production technologies for many marine ornamental species are emerging recent years, it can be reasonably predicted that the percentage of wild caught marine ornamental species will continue to dominate the sector on a global basis in the near future. Based on the current technologies it is neither possible nor economically viable to hatchery produce all the species required for the trade. Hence the hatchery production of marine ornamental species can be complementary but may not be full replacement for the collection from the wild. The idea behind the establishment of GMAD and MAC is indicative of the modern trend on wild collection of marine ornamental species. It shows that the wild collection sector is important and at the same time the protection of reefs due to wild collection also is vital. It also emphasizes the accurate data base needed, the ecofriendly methods to be followed and the commitment to a certified wild caught industry for sustainable trade. The attitude towards the wild caught sector and tank reared sector of aquarium industry should be mutually supportive. In the immediate future India can emerge as one of the major source countries for a sustainable marine ornamental trade if we appropriate steps in the light of the current global scenario by formulating appropriate policies for wild collection of species and also by commercial production of suitable species though the development of hatchery technologies.

References

Bertram, 1996. The aquarium fishery in the Cook Islands. Is there a need for management? *Secretariat of the Pacific Community Live reef Fish Information Bulletin*, 1: 10–12

Bunting, B., Holthus, P. and Spalding, S., 2003. The marine aquarium industry and reef conservation. In: *Marine Ornamental Species: Collection, Culture and Conservation*, (Eds.) J. Cato and C.Brown. Iowa State Press, Ames, USA, pp. 109–124.

Cervino, J., Hayes, R., Honovich, M., Gorea, T., Johns, S. and Rubec, P., 2003. Changes in zooxanthellae diversity, morphology and miotic index in hermatypic corals and anemones exposed to cyanide. *Mar. Poll. Bull.*, 46: 573–586.

Colette, W., Taylor, M., Green, E. and Razak, T., 2003. *From Ocean to Aquarium: A Global Trade in Marine Ornamental Species*. UNEP World conservation and monitoring centre (WCMC), 65 pp.

Corbin, J. and Young, L., 1995. *Growing the Aquarium Products Industry for Hawaii*. Dept. of Land and Natural Resources Aquaculture Development Programme, Hawaii, 35 pp.

Edwards, A. and Shepherd, A., 1992. Environmental implications of aquarium fish collection in the Maldives with proposals for regulation. *Environmental Conservation*, 19: 61–72.

Friedlander, A., 2001. Essential fish habitat and effective design of marine reserves: Application for marine ornamental fishes. *Aquarium Science and Conservation*, 3: 135–150.

Gopakumar, G., 2007. Diversity and conservation of marine ornamental gishes. In: *Biodiversity in India: Issues and Concerns*, (Eds.) S. Kannaiyan and A. Gopalam. Associated Publishing Co., New Delhi, p. 34–44.

Gopakumar, G., George, Rani Mary and Jasmine, S., 2001. Hatchery production of the clown fish *Amphiprion chrysogaster*. In: *Perspectives in Mariculture*, (Eds.) N.G. Menon and P.P. Pillai. Marine Biological Association of India, Kochi, p. 305–310.

Gopakumar, G., Sriraj, G., Ajithkumar, T.T., Sukumaran, T.N., Raju, B., Unnikrishnan, C., Hillari, P. and Benziger, V.P., 2002.

Breeding and larval rearing of three species of damsel fishes (Family Pomacentridae). *Mar. Fish. Infor. Ser.* (*T&E*), 171: 3–5.

Gopakumar, G. and Santhosi, I., 2009. Use of copepods as live feed for larviculture of damselfishes. *Asian Fisheries Science*, 22: 1–6.

Gopakumar, G., 2008a. Resource analysis, trade potential and conservation and management of marine ornamentals in India. In: *Ornamental Fish Breeding: Farming and Trade*, (Eds.) B.M. Kurup, M.R. Boopendranath, K. Ravindran, Saira Banu and A.G. Nair. Dept. of Fisheries, Govt. of Kerala, India, p. 64–79.

Gopakumar, G., 2008b. Trends in the development of marine ornamental fish trade: An Indian perspective with emphasis on status of breeding technology and management policies for long term sustainability. *International Seminar on Ornamental Fish Breeding: Farming and Trade*, Dept. of Fisheries, Govt. of Kerala, Book of Abstracts, p. 11.

Gopakumar, G., Santhosi, I. and Ramamurthy, N., 2009. Breeding and larviculture of sapphire devil damsel fish *Chrysiptera cyanea. J. Mar. Biol. Ass.*, India, 51(2): 130–136.

Gopakumar, G., Ignatius, Boby, Santhosi, I. and Ramamoorthy, N., 2009. Controlled breeding and larval rearing techniques of marine ornamental fishes. *Asian Fisheries Science*, 22: 787–804.

Gopakumar, G., Madhu, K., Madhu, Rema and Ignatius, Boby, 2008. Hatchery production of marine ornamental fishes as an alternative for indiscriminate exploitation from coral reef habitats. In: *International Conference on Biodiversity Conservation and Management*, Book of Abstracts, pp. 163–164.

Gopakumar, G., Santhosi, I. and Ramamoorthy, N., 2009. Breeding and larviculture of the sapphire devil damsel fish *Chrysiptera cyanea*. In: *MECOS 2009*, Book of Abstracts–Marine Biological Association of India on Feb 9–12, Cochin, India, pp. 176–177.

Harriot, V., 2003. Can corals be harvested substantially? *Ambio*, 32: 130–133.

Hestinga, G., Perron, F. and Orako, 1984. Mass culture of giant clams (Family Tridacnidae) in Palau. *Aquaculture*, 39: 197–215.

Hutchings, J., 2002. Life histories of fish. In: *Handbook of Fish and Fisheries, Vol. 1*, (Eds.) P. Hart and J. Reynolds, p. 149–174.

Larkin, S. and Degener, R., 2001. The US wholesale market for marine ornamentals. *Aquarium Science and Conservation*, 31(1–3): 13–24.

Lubbock, H. and Polunin, 1975. Conservation and the tropical marine aquarium trade. *Environmental Conservation*, 2: 229–232.

Madhu, K. and Rema, M., 2002. Successful breeding of common clown fish under captive conditions in Andaman and Nicobar Islands. *Fishing Chimes*, 22(9): 16–17.

Madhu, K. and Rema, M., 2002. Successful breeding of common clown fish under captive conditions in Andaman and Nicobar Islands. *Fishing Chimes*, 22(9): 16–17.

Madhu, K., Madhu, Rema, Gopakumar, G., Rajagopalan, M. Krishnan, L. and Ignatius, Boby, 2008. Captive breeding and seed production of marine ornamental fishes of India. In: *Ornamental Fish Breeding: Farming and Trade*, (Eds.) B.M. Kurup, M.R. Boopendranath, K. Ravindran, Saira Banu and A.G. Nair. Dept. of Fisheries, Govt. of Kerala, India, p. 142–146.

Ogawa, T. and Brown, C., 2001. Ornamental fish aquaculture and collection in Hawaii. *Aquarium Sciences and Conservation*, 3: 151–169.

Oliver, K., 2003. World trade in ornamental species. In: *Marine Ornamental Species: Collection, Culture and Conservation*, (Eds.) J. Cato and C. Brown, p. 49–63.

Pananghat, Vijayagopal, Gopakumar, G. and Vijayan, Koyadan Kizhakedath, 2008. Empirical feed formulations for the marine ornamental fish, striped damsel, *Dascyllus aruanus* (Linne´ 1758) and their physical, chemical and nutritional evaluation. *Aquaculture Research*, 39: 1658–1665.

Rema, Madhu, Madhu, K. and Gopakumar, G., 2007. Broodstock development and captive breeding of maroon clown, *Premnas biaculeatus*. In: *Fisheries and Aquaculture: Strategic Outlook for Asia*, Book of Abstracts–8th Asian Fisheries Forum (organised by Asian Fisheries Society, Indian Branch, Nov. 20–23, Kochi, India, p. 148.

Rema Madhu, Madhu, K., Gopakumar, G., Rajagopalan, M., Krishnan, L. and Ignatius, Boby, 2008. Larvi-feed culture for seed production of Ornamental fishes. In: *Ornamental Fish*

Breeding: Farming and Trade, (Eds.) B.M. Kurup, M.R. Boopendranath, K. Ravindran, Saira Banu and A.G. Nair. Dept. of Fisheries, Govt. of Kerala, India, p. 147–154.

Rubec, P., 1987. Fish capture methods and Philippine coral reefs. *IMA Philippines visit. Part II, Marine Fish Monitor,* 2(7): 30–31.

Soegiarto, A. and Polunin, N., 1982. *The Marine Environment of Indonesia.* Government of Indonesia under sponsorship of IUCN and WWF.

Tissot, B., 1999. Adaptive management of aquarium fish collection in Hawaii. *Secretariat of Pacific Community, Live Reef Fish Information Bulletin,* 6: 16–19.

Tissot, B. and Hallacher, L., 1999. *Impacts of Aquarium Contractors on Coral Reef Fishes in Kona, Hawaii.* Department of Land and Natural Resources, Division of Aquatic Resources, Honolulu, Hawaii, USA.

Vallejo, 1997. Survey and review of the Philippine marine aquarium fishing industry. *Sea Wind,* 11: 2–16.

Wood, E., 1985. *Exploitation of Coral Reef Fishes for Aquarium Trade.* Marine Conservation Society, Ross-on-Wye, U.K., pp. 121.

Wood, E., 2001. *Collection of Coral Reef Fish for Aquaria: Global Trade, Conservation Issues and Management Strategies.* Marine Conservation Society, UK, 80 pp.

Chapter 11

Marine Ornamental Fishery Resource and its Management at Vizhinjam Coast, Southern Kerala

M.K. Anil, S. Jasmine, B. Santhosh, Rani Mary George, B. Raju, C. Unnikrishnan and H. Jose Kingsly

Research Centre of the Central Marine Fisheries Research Institute, Vizhinjam, Kerala – 695 221

Reef fishes with their brilliant colours and morphological peculiarities have always fascinated marine aquarium keepers. It is estimated that 1.5–2.0 million people worldwide keep marine aquaria, and the value of annual marine ornamental trade is estimated to range between US$ 200–330 million. More than 98 per cent of the marine ornamental fishes marketed are wild collected from oral reefs of tropical countries. Vizhinjam coast is unique with its stable water quality and rocky bottom having patchy coral reef. It harbours a rich variety of vibrantly coloured marine ornamentals. This fauna includes more than 150 species, which include marine angels (Family: Pomacanthidae), Butterfly and Banner fishes

(Chaetodontidae), Moorish idol (Xanclidae), Batfishes (Ephippidae), Trigger and file fishes (Ballistidae), Damsels (Pomacentridae), Surgeon fishes (Acanthuridae), Rabbit fishes (Siganidae), Squirrels (Holocentridae), Porcupine fishes (Diodontidae), Puffer fishes (Triodontidae) Marine eels (Anguillidae), Cardinal fishes (Apogonidae), Groupers (Serranidae), Snappers (Lutjanidae), Cow fish and Box fish (Ostraciidae), Razor fish, Wrasses, Hog fish, (Labridae) and Parrot fishes (Scaridae), Pipe fish and sea horse (Syngnathidae), Sucker fish (Echeneidae), Sweet lips (Haemulidae) etc.

Marine ornamental fishes are distributed chiefly in tropical seas particularly in the coral reef habitats and also in regions which have rocky bottom. More than 4,000 species of fish live on coral reefs and associated habitats of Indo-Pacific. Marine aquarium hobby has grown rapidly and steadily in recent years, mainly due to the scientific advancements marine aquarium and reef tank technology and has resulted in proportionate expansion in global marine ornamental fish trade. It is estimated that 1.5–2.0 million people worldwide keep marine aquaria, and the value of annual marine ornamental trade is estimated to range between US$ 200–330 million. Nearly all the tropical marine aquarium fish and invertebrate in trade are taken from or around coral reefs. Only 25 species are cultured on a commercial basis but the bulk of specimens, probably more than 98 per cent, are taken from the wild.

Aquarium collectors normally catch non-food species and are economically valuable, which should provide an incentive to conserve coral reef ecosystems. This industry provides jobs and income for many people, particularly in low income groups such as fishermen and to those involved in holding, packing and transportation. Destructive and non selective methods used to capture fish raises conservation issues, and there has been a long-running debate about the positive and negative aspects of collecting ornamental fishes from wild. The main issues are possible over-exploitation of selected species, secondly effects of this on reef inhabitants, damaging methods of collection and high post-harvest mortalities.

Methods Used for Collecting Marine Ornamental Fishes

Fish Traps

Fish traps are made of 6 to 10 mm MS rod frame work covered with chicken mesh and there is a vestibule made at an angle so that fish which enters the trap will not escape easily. There is a small window at the rear end which can be opened to collect the fishes from the trap. Crushed mussel is put in the raft as bait and the cage can be lifted after 5-6 hr but usually the next day morning.

Lift Traps

It is a rectangular or square shaped box without a lid made of ms rod and covered with 10 mm mesh netting/mosquito cloth. It is baited with crushed mussels or fishes and placed in rocky or coral areas. They are lifted at regular intervals according to the availability of fishes.

Dragnets and Skin Diving

Two types of drag nets are commonly used for marine ornamental fish collection.

1. A large net of 20-30 mm mesh size (1/2 No. netting) of 50-60 m (2.5 kg) length with foot rope with lead weight (30 g) at 15 cm intervals and head rope with floats at 30 cm intervals and is used for catching medium type fishes. Usually a motorized or manual catamaran and two to three persons including a diver is required for the operation to surround and trap fish in the net. Diver will net the fish from the trap using scoop net while the other person will be controlling the catamaran which will also contain buckets or containers with battery operated aerator for keeping the fishes caught and to transport them to shore.
2. Rectangular net made of mosquito cloth or 8 mm mesh nylon net of 10 to 20 m length is used to surround damsel shoals near rocks or other structures. Two persons with atleast one diver are required for this operation. Floats and sinkers are fixed same as the above type. The trapped fishes are scooped out.

Hook and Line

Hook and line is used to catch some ornamental groupers and eels.

Skin/SCUBA Diving

Slow moving fishes like sea horse, pipe fish, trigger fish, bat fish and lion fish are easily caught using a scoop net.

By-catch

Fishes collected from shore seine, boat seine and lobster trap.

Methods of Collection *vs* Species

Different type of gears used for the collection resulted in different groups of fishes though some fishes are caught using different nets. The fishes caught in fish traps include butterfly and banner fishes (Chaetodontidae), moorish idol (Xanclidae), trigger and file fishes (Ballistidae), damsels - larger species (Pomacentridae), surgeon fishes (Acanthuridae), rabbit fishes (Siganidae), squirrels (Holocentridae), puffer fishes (Triodontidae) marine eels (Anguillidae), cardinal fishes (Apogonidae), groupers and soap fishes (Serranidae), snappers (Lutjanidae), cow fish and box fish (Ostraciidae), razor fish, wrasses, hog fish, (Labridae) parrot fishes (Scaridae) and sweet lips (Haemulidae).

Catch from lift trap and small meshed drag net are mostly small fishes (yellowtail and black damsels and surgeon major of the family (Pomacentridae). The fishes caught in dragnet of mesh size 30 mm and above include marine angels (Family: Pomacanthidae), butterfly and banner fishes, moorish idol, batfishes (Ephippidae), trigger and file fishes, damsels, Surgeon fishes, rabbit fishes, squirrels, porcupine fishes, puffer fishes, cardinal fishes, snappers, cow fish and box fish, razor fish, wrasses, hog fish, and parrot fishes, pipe fish (Syngnathidae), sweet lips etc. Hook and line is used for catching ornamental groupers like *Cephalopholis sonnerati* and *C. argus, Epinephelus merra* and eels. Eels are also caught in traps when fish pieces are used as bait. Skin and scuba divers collect angels, butterfly fishes, lion fishes and scorpion fishes (Scorpaenidae), pipe fish and sea horse (Syngnathidae). By catch from conventional fishing nets such as shore seine, boat seine and lobster trap include porcupine fishes, puffer fishes, cow fish, groupers and sucker fish (Echeneidae), etc.

The capacity to collect fish without damaging the fragile reef requires considerable skill and experience. A number of techniques are used for collecting fishes without injury. In Sri Lanka, some collectors use specially made small, tubular nets for capturing species that live in shallow burrows. The mouth of the net is placed over the entrance to the burrow, and the fish are forced out of their refuge using a fine rod. As they emerge into the net, the mouth is closed with a drawstring. In Sri Lanka and the Maldives most collectors capture the bulk of their specimens with a combination of large and small hand nets. A few collectors in Hawaii use a seine-like net which is less than 2 m long, weighted at the bottom and has a wooden pole at each end. The collector sets one pole down then encircles the fish with the other. This is particularly effective for catching certain species, such as the fire- fish (Randall, 1987). Collection needs to be properly managed, so that species are not over- exploited, and only those suitable for captivity are to be fished out. Steps also need to be taken to make sure that reef habitat is not damaged during fishing.

Fishes Caught along the Vizhinjam Coast

One hundred and forty eight species belonging to 31 families were caught during the survey.

Acanthuridae (Surgeon Fishes and Tangs)

Many of the species are brightly colored and popular for aquaria. The distinctive characteristic of the family is the spines, one or more on either side of the tail, which are dangerously sharp like the ones used by surgeons and hence the name. The small mouths have a single row of teeth used for grazing on algae.

A very popular fish with hobbyist, but one that is not easy to care for, is the Powder Blue Tang Acanthurus leucosternon. This gorgeous fish is sky blue overall, with a yellow dorsal fin, a white anal fin, a black head and a white band behind the head. It is perhaps the most beautiful member of this genus. Other fishes of this group include lined surgeon fish *A. lineatus*, convict surgeon fish *A. triostegus*, elongate surgeon fish *A. mata* and chocolate surgeon fish *A. pyroferus*, *A. xanthopterus* and *Naso annulatus*. Surgeon fishes are seen throughout the year with peak in April and August but *A. Leucosternon* and *N. annulatus* are not common.

Antennariidae (Frogfishes)

Frogfishes of the family Antennariidae are generally small, globose anglerfishes easily distinguished from members of related families by the presence of three well-developed dorsal spines, laterally directed eyes, a large, anterodorsally directed mouth, and a short, laterally compressed body. Species available in Vizhinjam area include *Histrio histrio* and *Antennarius nummifer.*

Apogonidae (Cardinal Fishes)

Apogonidae is one of the largest tropical reef fish families. These are small (usually under 10 cm), laterally compressed, often brightly coloured fishes. They have two separate dorsal fins; II anal spines; large eyes; a moderately large oblique mouth; and preopercle with a ridge in front the margin. Important ornamental species include *Archamia fucata, Archamia lineolata, Apogon endekataenia, Apogon taeniatus, Apogon nigrofasciatus, Apogon fasciatus, Apogon aureus* and *Apogon psuedotaeniatus.* They were encountered throughout the year with peaks in February and September.

Balistidae (Trigger Fishes)

Triggerfish have an oval, highly compressed body. The head is large, terminated in a small but strong- jawed mouth with teeth adapted for crushing shells. The eyes are small, set far back from the mouth, at the top of the head. The anterior dorsal fin is reduced to a set of three spines. The first spine is stout and by far the longest. All three are normally retracted into a groove. The sickle shaped caudal fin is used only to escape predators. The important members having ornamental potential include *Balistapus undulates, Odonus niger, Sufflamen fraenatus, Psuedoballistes flavimarginatus, Abalistes stellatus, Canthidermis maculate* and *Psuedoballistes fuscus.* Commonly seen during monsoon and post monsoon months with peak in July–August.

Caesionidae (Fusilier Fishes)

They are cylindrical, sleek fishes related to the snappers and are planktivorus. *Pterocaesio chrysozona* is an ornamental caesionid available in Vizhinjam coast but not a hardy species for novice.

Carangidae (Scads)

Carangidae is a family of fish which includes the jacks, pompanos, jack mackerels and scads. Body generally compressed, although body shape extremely variable from very deep to fusiform. Most species are seen with only small cycloid scales. Scales along lateral line often modified into spiny scutes. Detached finlets, as many as nine, sometimes found behind dorsal and anal fins. Large juveniles and adults with 2 dorsal fins. Anal spines usually 3, the first 2 separate from the rest. Widely forked caudal fin. Avery colourful member of this family from the area is *Gnathanodon speciosus*.

Chaetodontidae (Butterfly Fishes)

Butterfly fishes with their amazing array of colors and patterns are among the most varied fishes of the aquarium, the majority of which live on or close to coral reefs. Most species measure from 5-9.5" (13-24 cm) in length and have deep, flattened bodies that are frequently adorned by extended fins.

The members of this family from this area include red tail butterfly *Chaetodon collare*, sunburst butterfly *C. kleinii*, Indian vagabond butterfly fish *C. decussatus*, yellow head butterflyfish *C. xanthocephalus*, threadfin butterfly fish *C. auriga*, gardner's butterfly fish *C. gardiner*, Raccoon butterfly fish *C. lunula*, speckled butterfly fish *C. citrinellus*, lined butterfly fish *C. lineolatus* Vagabond butterfly fish *C. vagabundus*, black-back butterfly fish *C. melannotus*, eight banded butterflyfish *C. octofasciatus* and blue blotch butterfly fish *C. plebeius*, melon butterflyfish *C. trifasciatus, C. selene* and the long nose butterfly fish *Forcipiger flavissimus*. Bannerfishes are also included in this group - their common name in German is "Wimplefish" meaning "Pennantfish" or "banner fish". Awimple is a type of hat with feathers of which the heightened dorsal is reminiscent. Banner fishes of the marine aquarium include *H. acuminatus, H. monoceros, H. pleurotaenia H. singularius, H. varius* and *Heniochus pleurotaenia*. Butterfly fishes are seen throughout the year with peak in April to September.

Diodontidae (Porcupine Fish)

A small family of marine, warm-to temperate-water fish in which the short, round body is covered by numerous hard spines. The teeth are fused in each jaw, resembling the beak of a arrot. Porcupine fish are well known for their peculiar habit of distending themselves with water until the body assumes a nearly spherical shape. They include *Diodon histrix, Diodon liturosus* and *Lophodiodon calori*. They are caught throughout the year in shore seines.

Figure 11.1: Marine Ornamental Fishes of Vizhinjam Coast

Echeneidae (Remoras or suckerfish)

Their first dorsal fin takes the form of a modified oval sucker-like organ that open and close to create suction and take a rigid hold against the skin of larger marine animals. Remoras are primarily tropical open-ocean dwellers, occasionally in coastal waters. Remoras are commonly found attached to sharks, manta rays, whales, turtles, and dugong (hence the common names sharksucker and whalesucker). Smaller remoras also fasten onto fish like tuna and swordfish, and some small remoras travel in the mouths or gills of large manta rays, ocean sunfish, swordfish, and sailfish. Echeneis naucrates is usually caught entangled in gill nets and they thrive well in aquarium tanks.

Haemulidae (Grunts)

The grunts are a family of small to medium-sized fishes that occur worldwide in tropical and temperate seas. The common name is derived from their habit of emitting grunting noises that results from grinding the upper and lower pharyngeal teeth. They are bottom-feeding predators eg. *Plectorhincus lineatus* and *P.vittatus*. They are not very common.

Holocentridae (Squirrel Fishes)

The members of this squirrelfish family are reddish in color mixed with silver and white; all have large eyes, and are nocturnal, hiding in crevices or beneath ledges. These are shallow water fishes, found from the surface to about 100 meters. Squirrel fishes have large, sharp, extremely rough scales. Members of this family include *Sargocentron rubrum, S. diadaema, S. spiniferum, Myripristis adusta, M. Murdjan, M. melanostictus, M. hexagonatus* and *Neoniphon samara*. They are caught throughout the year with a peak in June.

Labridae (Wrasses)

They are very beautiful and real funny fishes for the marine aquarium. Wrasses come in a wide assortment of colors, shapes and sizes. Some of the beautiful members of this family in the aquarium are *Thalassoma jansenii* (Jansen's wrasse), *T. lunare* (Moon wrasse), *Halichoeres zeylonicus* (Gold stripe wrasse), *H. melanochir* (Yellowtailed wrasse), *Anampses liniatus* (Lined wrasse), *Labroides dimidatus, Hemigymnus melapterus, Coris formosa, Cheilinus chlorouru, Bodianus neilli, Xirichtys novacula* and *Thalassoma hardwicke.*

Lutjanidae (Snappers)

Snappers are a prominent family of predatory, small to medium-sized fishes with an ovate to elongate, moderately compressed body. Most snappers dwell in shallow to intermediate depths in the vicinity of reefs, although there are some species largely confined to depths between 100 and 500m. Alarge number of species have "snapper" in their common name; most but not all are Lutjanidae. Ones which are with ornamental qualities include *Lutjanus lutjanus, L. fulvus, L. quinquilineatus, L. biguttatus, L. russelli, L. decussatus, L. bohar, L.vitta* and *L. fulviflammus.*

Monodactylidae (Moony Fishes)

Fishes of this family are commonly referred to as monos or moonyfishes. All are laterally compressed with a round disc-shaped body. There are scales on the dorsal fin. They are silvery with yellow and black markings; and are popular as aquarium fish. *Monodactylus argenteus* is commonly encountered in Vizhinjam area throughout the year.

Monacanthidae (Filefish)

Filefish (also known as foolfish, leatherjackets or shingles) are close relatives of triggerfishes (Balistidae), but differ from them by having more compressed bodies, a more pointed snout, a longer first dorsal spine, a very small second dorsal spine (sometimes absent), and no third dorsal spine. Members of this family encountered along this coast include *Cantherhines sandwichiensis* and *Pervagor alternans.*

Mullidae (Goatfishes)

Many species of goatfish are conspicuously colored; however, they are not popular in aquaria as they will not easily adjust with captive conditions. Members of this family include *Upenaeus moluccensis, U. tragula, U. sulphureus* and *Parupaenaeus indicus.*

Muraenidae (Eels)

Moray eels are serpentine in appearance and the dorsal fin extends from just behind the head along the back and joins flawlessly with the caudal and anal fins and most species ack pectoral and pelvic fins. They prefer to hide in the cave among rocks and coral reef.

Species of moray eels in the coast include *Gymnothorax flavimarginatus* (Yellow edged moray), *G. javanicus* (Giant moray), *Echidna nebulosa* (Snowflake moray), *Gymnothorax javanicus, G. favagineus, G. zonipectus, G. meleagris* and *G. flavimarginatus.*

Nemipteridae (Threadfin Breams)

They are also known as whiptail breams and false snappers. They are found in tropical waters of the Indian and western Pacific

Oceans. Most species are benthic (bottom-feeding), carnivorous, eating small fishes, cephalopods, crustaceans and polychaetes; however, a few species eat plankton. *Scolopsis vosmeri* and *S. eriomma* can be acclimated to tank conditions and are caught in drag net collections.

Ostraciidae (Box and Cow Fishes)

Boxfishes Ostracion cubicus are comical little fishes that really do look like a box. They have a stiff cube shaped body that is covered with a hard bony armor. This provides excellent protection for these slow swimming fishes. Some types of Boxfish have "horns" on their head and thus they are also known as Cowfish (*Lactaria cornutus*). Box fishes are very common in shallow waters during monsoon (SW) and post monsoon moths with peak in September. Cow fishes are not common in this coast.

Pempheridae (Sweeper Fish)

Pempheridae are commonly called as sweeper, bullseye they have a fairly deep body, a short dorsal fin, a long-based anal fin, and large eyes. They show schooling, shallow-water species found near caves, rock ledges, reefs etc. Species such as *Pempheris vanicolensis* and *P. oualensis* also were found in the coast.

Ephippidae (Bat Fishes)

They are compressed laterally and deep-bodied with small mouth and anal fin with 3 spines Omnivores–feeds on algae and small invertebrates. Fishes of Platax genus are popular and unproblematic aquarium species, but grow very fast. Species available along the coast include *Platax orbicularis, P. teira* and *P. pinnatus.*

Pomacanthidae (Angel Fishes)

Marine angelfishes with their vibrant colours and deep, laterally compressed bodies, marine angelfishes are some of the more conspicuous residents of the aquarium. They are distinguished from butterfly fishes by the presence of strong preopercular spines in the former. This feature also explains the family name Pomacanthidae; from the Greek poma meaning "cover" and pakantha meaning "thorn". Common angels of the aquarium include koran angelfish Pomacanthus semicirculatus which has Arabic like inscriptions on its tail emperor angelfish *P. imperator*, blue-ring angelfish *P. annularis* Indian yellow-tail angelfish *Apolemichthys xanthurus* and coral beauty angel *Centropyge flavipectoralis*.

Pomacentridae (Damsels)

Damsels, clown fishes and sergeant majors belong to the family Pomacentridae and members of two genera namely Amphiprion and Premnas are commonly called as clown fishes.

Most damselfishes are extremely hardy, colorful and lively. Colourful ones in the area include *Pomacentrus caeruleus, P. pavo, P. tripunctatus, Neopomacentrus nemurus, N. filamentosus, Dascyllus trimaculatus* and *Chrysiptera unimaculata*. The Sergeant Major or píntanos are colourful damselfishes. It earns its name from its brightly striped sides, which are reminiscent of the insignia of a military Sergeant Major. The other species include *Abudefduf saxatilis, A. sordidus* and *A. bengalensis*.

Scaridae (Parrot Fishes)

Mostly a tropical fish group with large cycloid scales and teeth fused or parrot like. They are herbivorous, usually scraping algae from dead coral substrates. Bits of rock eaten with the algae are crushed into sand and ground with the algae to aid in digestion. *Chlorurus sordidus* is very common in this area seen throughout the year, peaking in March to May and October to December.

Scatophagidae (Scats)

A coastal fish family, with a deep body, two separate dorsal fins, and, typically, four spines in the anal fin. Scatophagus argus of this family could be used both in marine and brackish water tanks.

Scorpaenidae (Lionfishes)

Lionfishes are the venomous marine fishes in the genera *Pterois, Parapterois Brachypterois Ebosia* or *Dendrochirus*, of the family Scorpaenidae. They are notable for their extremely long and separated spines, and have a generally striped appearance. Species available in the study area include red lion fish *Pterois volitans*, broad barbed fire fish *P. antennata*, zebra turkey fish *Dendrochirus*, zebra black foot fire fish *Parapterois heterura, Humpback scorpionfish* and *Scorpaenopsis gibbosa*.

The venom of the lionfish, delivered via an array of up to 18 needle-like dorsal fins, is purely defensive. A sting from a lionfish is extremely painful to humans and can cause nausea and breathing difficulties, but is rarely fatal.

Serranidae (Groupers)

This family contains about 450 species of belonging 64 genera, including the sea basses and the groupers of which many are brightly colored. They are usually found over reefs and rocky coastal areas. Members of the family with ornamental value include *Cephalopholis sonnerati, Cephalopholis argus, Epinephelus merra, E. caeruleopunctatus, E. Malabaricus, E. tauvina* and *Grammistes sexlineatus.*

Siganidae (Rabbitfishes)

The Siganidae commonly known as rabbitfishes, from their peaceful nature, rounded snout, and rabbit-like appearance of the mouth. They are herbivores fishes that browse individually or in schools or feed on plankton. Species available in the region include *Siganus javus, S. canaliculatus* and *S. punctatus.*

Sygnathidae (Seahorse and Pipe Fish)

Common species belonging to this area include seahorse *Hippocampus kuda* and pipefish *Syngnathoides biaculeatus.* Seahorses have elongated head and snout, flexed at right angles to the body, which resembles a horse. They feed on minute organisms and protected by thin bony plates that are derivatives of the scales found in fishes. The seahorse swims weakly in an upright position by means of rapid, humming birdlike beats of its fins; at rest it curls its thin, prehensile tail around seaweed. One peculiarity of this fish is that the female forces the eggs into a pouch on the underside of the male, where they are fertilized and the male carries the young ones in the pouch and delivers the babies when they are ready to be released into the sea.

Theraponidae (Grunter, Tigerfish)

This is a family of marine and freshwater fish that have elongate to oblong, compressed bodies, spiny and soft-rayed parts of the dorsal fins separated by a distinct notch, a slightly forked tail fin and the pelvic fins inserted behind the bases of the pectoral fins. *Therapon jarbua* is usually caught in shore seines from coastal waters.

Tetraodontidae (Puffers, Globefish)

They are primarily marine or estuarine fishes. The scientific name, Tetraodontidae, refers to the four large teeth, fused into an upper and lower plate, which are used for crushing the shells of crustaceans and mollusks, and worms, their usual prey. Puffer fish are the second most poisonous vertebrate in the world, the first being a Golden Poison Frog. The skin and certain internal organs of many tetraodontidae are highly toxic to humans, but nevertheless the meat of some species is considered a delicacy in both Japan and Korea. Commonly caught species from the coast which include *Canthigaster solandri, Arathron hispidus, A. nigrofasciatus* and *A. immaculatus.*

Zanclidae (Moorish Idol)

Zanclus cornutus is the sole representative of the family Zanclidae (from the Greek zagkios, "oblique"). A common inhabitant of tropical to subtropical reefs and lagoons with characteristically compressed and disk-like bodies, moorish idols stand out in contrasting bands of black, white and yellow which make them look very attractive to aquarium keepers. Moorish idol got its name from the Moors of Africa, who purportedly believe the fish to bring of happiness. Moorish idols are also popular aquarium fish, but despite their popularity, they are notorious for their short aquarium lifespans and difficulty. The fish have relatively small fins, except for the dorsal fin whose 6 or 7 spines are dramatically elongated to form a trailing, sickle-shaped crest called the philomantis extension. Moorish idols have small terminal mouths at the end of long, tubular snouts; many long bristle-like teeth line the mouth.

Marine ornamental fishes are usually collected from the wild. The fishes have to be collected by eco-friendly methods which do not damage the coral reef ecosystem. The collection methods should avoid any injury to the fishes. The ideal gears for the collection fishes from the wild are bag nets, traps and scoop nets. Bag nets are ideal to collect damselfishes, cleaner wrasse, cat fishes, cardinal fishes, bat fishes, porcupine fishes and box fishes. Traps can be placed in the reef with suitable baits. The ornamental fishes trapped inside can be collected at frequent intervals. The slow swimming fishes can be collected by diving and scoop netting. There are a number of measures that can be taken to make sure that ornamental resources and the habitats they come from are conserved and sustainably managed and it is essential that effective ornamental fishery management plans and regulations are developed and enforced. The management measures fall into four main categories: fishery management measures; improved industry standards; development of alternatives to wild harvest; and international trade restrictions. Ways to ensure that collection is maintained at a sustainable level should include a combination of management strategies implemented through an adaptive process, such as a limit on the total number of collectors, quotas developed for each species based on their life history traits and their abundance in the proposed collection area; maximum and minimum sizes for collection; and restrictions on the collection of rare species.

Habitat damage and negative impact to the ecosystem have to be avoided. Species that are incompatible to aquaria should be avoided. The establishment of marine reserves where the collection of marine ornamentals are made illegal, setting up of quotas and size limits, temporary closures and restricting access to the ornamental fishery through the use of permits can also reduce exploitation pressure. Pressure can be taken off wild populations by supplying maricultured ones, rather than wild-caught, fish for the marine aquarium market. It is the only answer to a long term sustainable trade of marine ornamentals. In addition hatchery produced fish are hardier and better adapted for captive rearing. Over 100 species of marine fish have been bred in captivity in many countries, but of these, relatively few have been bred in commercial quantities (Dawes, 1999). Marine fishes are more difficult to raise successfully in captivity than freshwater or brackish ones. Rearing those with pelagic eggs is particularly difficult, and there is more chance of success with species that have demersal eggs. More attention has to be paid to larval diet, which requires research into appropriate microalgae and zooplankton. Feeds enriched with essential elements are proving to be the key to success in many cases. Once the larvae have metamorphosed, mortality drops and the juveniles are usually relatively hardy. Management of marine ornamental fisheries is undeniably very complex but need to be managed to ensure they are biologically sustainable.

References

Bruckner, A.W., 2005. The importance of the marine ornamental reef fish trade in the wider. *Caribbean Rev. Biol. Trop.*, p. 127–138.

Dawes, J., 1999. International experience in ornamental marine species management–Part 2: some resource management strategies. *OFI Journal,* 27: 10–12.

Moe, M.A., 1999. *Marine Ornamental Aquaculture.* First International Conference of Marine Ornamentals, Hawaii.

Randall, J.E., 1987. Collecting reef fishes for aquaria. In: *Human Impacts on Coral Reefs: Facts and Recommendations,* (Ed.) B. Salvat. Antenne Museum E.P.H.E., French Polynesia, p. 29–39.

Wood, E.M., 1985. *Exploitation of Coral Reef Fishes for the Aquarium Trade.* Report to the Marine Conservation Society, 121 pp.

Wood, E.M. and Rajasuriya, A., 1996. *Handbook of Protected Marine Species in Sri Lanka.* Marine Conservation Society and National Aquatic Resources Agency, 25 pp.

Chapter 12

Techniques for Mass Production of Two Species of Clown Fish: Clown Anemone Fish *Amphiprion ocellaris* (Cuvier, 1830) and Spinecheek Anemone Fish *Premnas biaculeatus* (Bloch, 1790)

M.K. Anil, B. Santhosh, S. Jasmine, Reenamole, G.R. Unnikrishnan, C. and A. Anukumar

Research Centre of the Central Marine Fisheries Research Institute, Vizhinjam, Kerala – 695 221

The marine ornamental fishery is a multi-million dollar industry that provides livelihood for thousands of fishers in developing countries and provides aquarium hobbyists with over 1400 species of marine fishes. This industry began in the 1930s in Sri Lanka as a small export fishery. More than 98 per cent of the marine ornamental fishes marketed even today are wild collected from coral reefs of tropical countries. Marine aquarium hobby has grown rapidly and steadily in recent years, mainly due to the scientific advancements marine aquarium and reef tank technology and has resulted in proportionate expansion in global marine ornamental fish trade. It is estimated that 1.5–2.0 million people worldwide keep marine aquaria, and the value of annual marine ornamental trade is estimated to range between US$ 200–330 million. Nearly all the tropical marine aquarium fish and invertebrate in trade are taken from or around coral reefs. Destructive and non selective methods used to capture fish raises conservation issues, and there has been a long-running debate about the positive and negative aspects of collecting ornamental fishes from wild. This pressure can be taken off wild populations by supplying tank-bred, rather than wild-caught, fish for the aquarium market. Marine fish have always been more difficult to raise successfully in captivity than freshwater ones. According to Dawes (1999) over 100 species of marine fish have been bred in captivity in many countries, but of these, relatively few have been bred in commercial quantities. Commercial culture began in 1972 but only about 25 species have so far fulfilled the three main requirements for commercial culture: high value, high demand and relative ease of culture in large numbers (Moe, 1999). The mainstay of this trade are anemonefish (*Amphiprion* spp. and *Premnas biaculeatus*). At present, much of the technological expertise is in the more prosperous consumer countries, and there are many companies investing in mariculture enterprises. Commercial culture is concentrated in the United States (especially Florida and Hawaii), Europe and Taiwan, with virtually none in the countries where stocks originate.

Establishment of mariculture facilities away from the countries of origin deprives these nations of income and puts people out of jobs. In the context of trade in ornamental species, source countries would be deprived of benefits if genetic resources were taken for breeding and sale elsewhere.

Clown fishes continue to be the most demanded marine tropical fish and the technologies available at present on marine ornamental fish breeding are mainly centered around clown fishes. There are 27 known clown fish species. They are distinguished and taxonomically separated from other damsel fish by their dependence on anemones for protection. They are further distinguished from other damsels by their large capsule - shaped eggs and large larvae at hatch. Indo-

West Pacific region contains 10 known species. Species with widely spread regions are *Amphiprion akallopisos* (Yellow skunk), *A. bicinctus, A. clarkii, A. chrysopterus* (orange anemonefish), *A. frenatus* (tomato clown fish), *A. melanopus* (fire clown fish), *A. ocellaris, A. perideraion* (Pink skunk), *A. rubrocinctus* (red anemonefish) and *A. sebae* (sebae clown). Anemonefish as their name defines, live in a mutualistic relationship with anemones. In nature, selection of preferred anemones is species specific. Primary benefits to clown fish from anemone association are protection of the pair, their nests and a protection of their progeny from predation. The fish achieves protection from stinging of anemones by means of the development of a special external mucus layer. Clown fish appear to be monogamous, pairing for life.

Clown fishes are popular, brightly coloured, hardy fishes with vibrant colors. The most common clown fish and the one that looks most like Nemo the world famous cartoon character- is the percula clown fish which is bright orange with white stripes Vizhinjam Research Centre of CMFRI has standardised the techniques for broodstock development, larval rearing and mass production techniques of two species of clown fishes. Clown anemonefish (*Amphiprion ocellaris*) can produce 200-500 eggs in a single spawning with an average of two spawning per month. Survival ranges from 40 to 75 per cent. The fishes can be marketed within four months period or these can be reared for an extended period of six months for better price. Spinecheek anemonefish (*Premnas biaculeatus*) can produce 200-800 eggs in a single spawning with an average two spawning per months. The survival ranges from 50-80 per cent. If a breeding pair is formed these may give continuous breeding for several years. The larvae of different fishes can be reared in a single tank. The present paper also deals with details of larval rearing and methods of mass scale production especially with minimal investment.

Brood Stock Development

Brood stock can be developed in FRP tanks, cement tanks or synthetic tanks. For smaller pairs 250-300 litre capacity tank will be sufficient. Tanks can be filled with 3/4th the capacity with seawater. A biological filter must be provided in the tank to keep ammonia and nitrite level under control.

Clown Fish (*Amphiprion ocellaris*) – Breeding Protocol

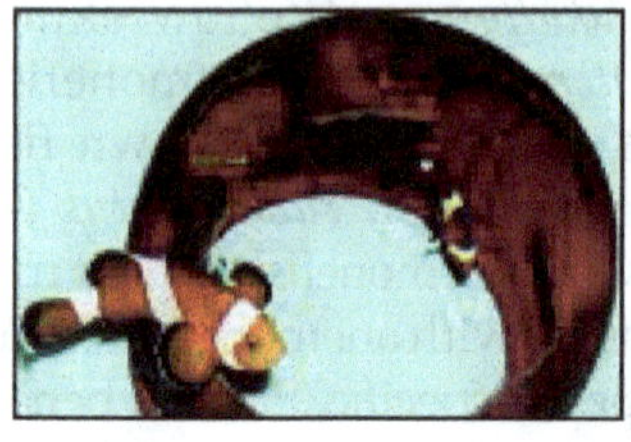

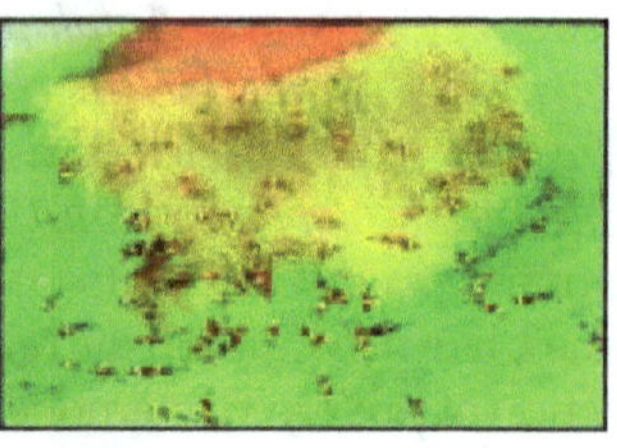

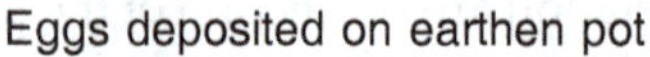

Eggs deposited on earthen pot | Hatchery produced juveniles

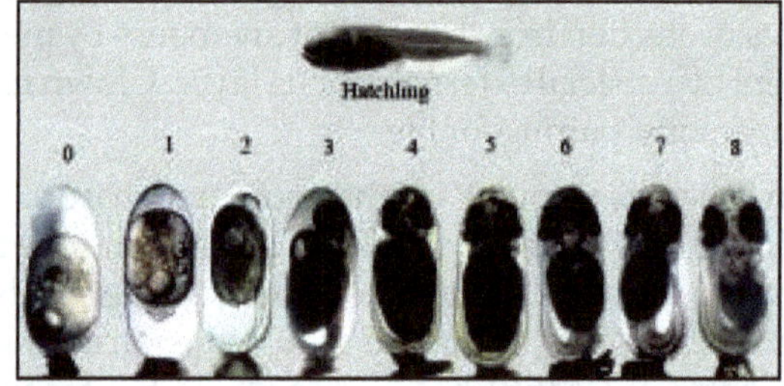

Development of embryo from 0-8 days of incubation

Clown Fish (*P.biaculeatus*) – Embroynic Development

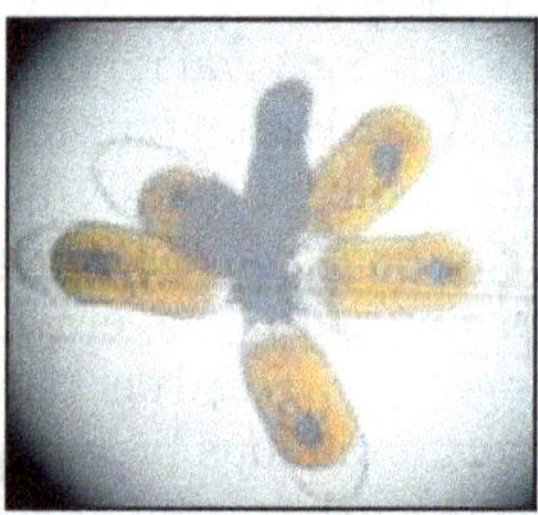

Eggs at the beginning of incubation

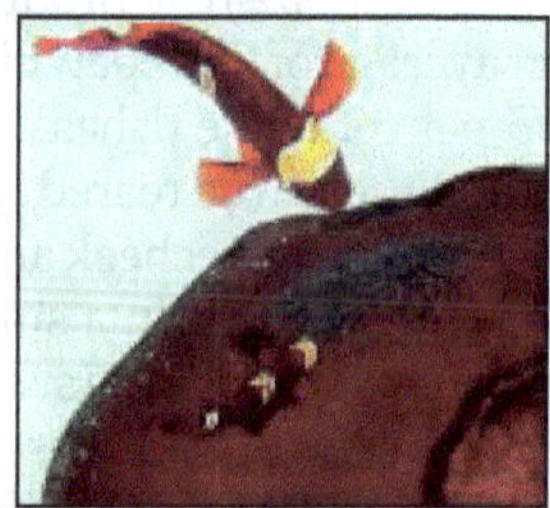

Silvery eggs on final day of incubation – *P. biaculeatus*

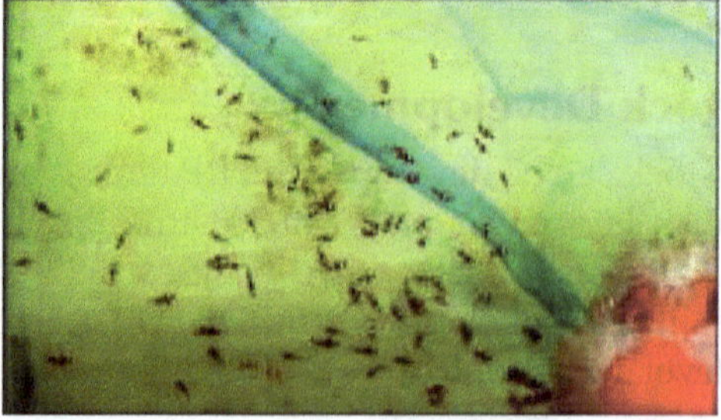

30 days old juveniles

Broodstock of clown fish can be obtained by collecting them from wild or purchasing healthy stock from reliable traders. Age of the fish is the most important factor determining sexual maturity. Sexually matured adult clown fish are usually 10-18 months old. While selecting possible pairs or purchasing fish for pairing, it is best to buy sub-adults. Even though they are younger than adult pairs, they will only take about 3-6 months for initial spawning. A distinct advantage when pairing clown fish is their ability to change sex. The best and easiest approach in pairing clown fish is purchasing 3 or 4 fish of equal size, 3.5 to 5 cm in total length. Put all the fish in one established tank with no other fish. Since sex reversal is common in clown fish, they simply decide which will become the male and which will become the female. Eventually, two fish will be moving together, chase others. Sometimes a pair will permit a few smaller individuals to remain as a reserve. Utilization of reserve fish is a unique adaptation in clown fish. When the female of a pair dies, the original male will become female and one of the reserve fishes will become the new male. Once pair formation has taken place they can be transferred to a separate breeding tank. Depending on the production capacity and seed demand several pairs can be maintained in a commercial hatchery. They can be stocked in a tank with or without a host anemone. If anemone is kept in the tank care must be taken to place the tank in such a position so as to have sufficient light to keep the anemone healthy.

Conditioning the Pair

Conditioning is manipulation of a combination of environmental factors to induce gonadal maturation and spawning. The factors may include light intensity, light duration, temperatre, water quality, nitrogen, phosphate, ammonia, pH, type of food, tank size and shape, aeration and habitat. Clown fish will mature and spawn within a wide range of temperatures from 21°C to 32° C. The salinity of around 28-32ppt is better while conditioning the fish. Lower salinity basically helps to reduce disease problems associated with parasites that demand higher salinities to survive. It also allows a large variance in salinity due to evaporation of tank water in the hatchery. A nitrate level of 20–30 ppm, nitrite and ammonia level of less than 0.1ppm, pH around 8.0–8.3 are ideal for conditioning tanks. Normally the clown fish utilize the live anemone as their protective habitat and the hard surface beneath the anemone as their spawning substrate. By creating an artificial environment in which a spawning pair are relaxed and feel protected, the anemone can be easily eliminated. Clown fish are territorial and will not accept interference from other clown fish or most other fish and invertebrates. For best results, the pair should be kept in individual tanks with opaque sides. Lighting should be moderate for the broodstock. Clown fish are basically bottom dwellers preferring some sort of habitat to hide in. A suitable spawning substrate like clay pot can be kept in the tank. They provide refuge for them and a suitable spawning substrate. In addition, they can be easily removed for hatching and cleaning and replaced quickly. Cleaning the spawning substrate should be regular.

Broodstock diets are nearly the main factor to gonadal maturation and successful spawning. Eggs contain considerable lipids which are energy resources needed for the prolonged development of the embryos within the eggs. These deposits are food reserves from mother deposited in the form of egg yolk. Hence nutritious diets in enough quantities must be fed to the broodstock fish. If broodstock fish are not suitably fed, the results are directly reflected in the number of eggs laid, fertilization rate, hatch rate and the quality of hatched larvae. Poor quality eggs develop slowly, hatch late and often result in significant early larval mortalities. Conditioning food should be given to the brood stock clown fish two times a day, in the morning and evening. If prepared brood stock diets are not available boiled and chopped mussel/clam/ prawn meat and fish roe can be fed libitum twice a day. Live feeds like copepods, mysids, *Artemia nauplii* and adult Artemia can also be supplemented. If brine shrimp or other live foods are included in the broodstock feed, they should be offered only as a supplement. A starved live animal is not a nutritionally balanced food. Essential fatty acids, micro algae, etc. can be administered to adult brine shrimp prior to feeding, to boost their food value.

Disturbance of the pairs in the broodstock tank must be minimal for more consistent spawning and fecundity. Disruption during the spawning of clown fish often results in scattered, not fully fertilized eggs. Walking through broodstock areas, can have an effect on some species. Any movement can cause the pair to temporarily stop spawning and move away from the new nest. It is better that every effort should be made to keep the pair isolated from external distractions. Routine daily procedures of broodstock maintenance include feeding, checking for new spawns, checking the general health of the pairs and nest and adjusting air, siphoning of detritus buildup and uneaten food. Monthly maintenance includes removal of dirty spawning substrates and replacing them without delay with clean ones. Every year, the entire tank should be thoroughly cleaned, gravel removed and new or reclaimed gravel replaced.

A day before spawning, the parents select the suitable site for laying eggs and start cleaning the area from algae and other attached animals. Spawning occurs during day time (6 AM -3.30 PM) and it lasts for about one to one and half hours. Each female lays 300 to 1000 capsule shaped eggs at every twelve to fifteen days interval. Generally the egg size ranges between 300- 400 micron in length. Each egg is attached to the substratum by a stalk. During the incubation period both the parents carefully look after the eggs by fanning the eggs by their fins and removing the dead and infected eggs by mouth. Newly spawned eggs are bright orange or pale pink in colour depending on species and these turned to black on 4-6th day and later to silvery colour with the eyes of the larvae prominent on the seventh day. The eggs hatch on the seventh day shortly after sunset at a water temperature range of 27–29° C.

Eggs can tolerate more mechanical and chemical changes than newly hatched larvae. While within the egg, larvae can adjust more easily to water changes than after hatch. The duration of hatching of clown fish eggs from the day of egg laying for the common species generally ranges from 6th day evening to 8th day evening, at a temperature range of 26-28° C. Hatching of clown fish eggs normally commences from 1-2 hours after dark. Hatching takes about 15-20 minutes. If larvae are allowed to remain in broodstock tanks overnight, numbers of larvae are significantly reduced due to predation or due to the filtration in the broodstock tanks. Scooping and siphoning hatched larvae is very impractical. Nets should never be used to transfer the larvae from spawning tank. It is important to realize that larvae cannot tolerate being touched by a solid object like a net. Common practice is to allow the larvae to hatch within the broodstock tank, drawn to a specific spot for removal by using a torch light. The accumulated larvae can be collected by using water filled buckets.

Hatching nests can be done outside the broodstock tanks also. This method is more advantageous especially to commercial operations. Larvae within the eggs are more tolerable than newly hatched larvae to physical, chemical and water quality change. Hatching nests within broodstock tanks may yield almost 100 per cent hatch, but recovery of the larvae will be less. Many larvae are consumed, drawn into the filters, become entrapped or die before being captured. Removing the nest and placing it into a flow-through hatching tank is better. To keep the eggs moving and well aerated, they were either aerated or incoming water flow was directed on to the eggs. Physically removing intact nests just prior to hatch and placing them directly in larval rearing tanks is also found to yield successful hatching.

On the expected day of hatching, larvae break their capsules and hatchlings emerge soon after sunset and peak hatching takes place between 7–8 pm. The newly hatched larvae measure 3-4 mm in length and each has transparent body, large eyes, visible mouth and a small yolk sac. Soon after hatching the larvae are free swimming. In the case of false clown *A. ocellaris* number of eggs per spawning ranged from 200 to 500 numbers and the normal spawning interval was 10 to 16 days. Hatching occurred on the evening of the 7 or 8th day of incubation. The newly hatched larvae measured from 340 to 400 µ. In the case of *Premnas biaculeatus*, the number of eggs per spawning ranged from 200 to 800 numbers and the spawning interval was 14 to 17 days. Hatching occurred on the evening of the 6-7th day of incubation. The newly hatched larvae measured from 340 to 400 µ.

Larval Rearing

Seawater used for larval rearing must be of very high quality. Water with less detritus and dissolved organic matter can be filtered through gravity sand filter and poured into rearing tanks using a filter bag of 0.2 micron. Water with high content of organic matter must be allowed to settle in a settling tank for two to three days. This water can be transferred to another tank where it is treated with chlorine at 10 ppm level overnight and then aerated to strip the residual chlorine. The chlorine content of the water must be checked before use and any excess chlorine must be removed using sodium thiosulphate. It is advisable to give a final filtration of stored water through a cartridge filter series of 10, 5, 2 and 0.2 micron before adding to larval rearing tank.

The larval rearing has to be carried out in green water and feeding with rotifers initially from 1-9 day post hatch (dph) and subsequently during 7 - 20 dph with newly hatched *Artemia nauplii.* A minimum 8-10 nos of rotifers per ml is required during rotifer feeding period and 2-5 nos nauplii per ml during Artemia feeding stage. The larvae metamorphose between 15-20 days. Artificial/ mocroencapsulated feed of proper size (100-200 micron) or pellets ground and sieved through different mesh can also be given as feed from 6-8 dph onwards. After metamorphosis the larvae can be transferred to grow out tanks with or without sea anemone. Mild aeration can be provided during larval rearing. The larviculture period from 3-8 dph is critical due to the complete dependence on exogenous feed. After 8 dph there will not be any further mortality if proper feeding and water quality parameters are maintained. The range of environmental parameters required are pH 8.0–8.2, temperature 26–30° C, DO 5.0–7.5 ml/lit, salinity 32–36 ppt, ammonia and nitrite at zero levels. The tank bottom should be cleaned daily. Minimum 25 per cent water has to be exchanged and sufficient greenwater should be added daily. Green water provide nutrition for rotifers in addition to improving water quality.

Water quality is the key environmental factor but can be easily controlled with simple water exchanges. Success in larval rearing is closely linked with availability of quantity and quality of live feeds and how they are administered. Without a ready, plentiful, nutritious live feed source your larval rearing attempts will be a failure. To rear clown fish larvae, about 300-600 rotifers per larvae per day for a period of 5-9 days are required. In addition, the rotifers must be completely nutritious and balanced with essential fatty acids and micro algae. It is necessary to clean the tank daily of detritus and uneaten food. Water exchanges and air flows must be checked. Larvae do not normally actively seek food but tend to be opportunistic feeders, patiently waiting for a food particle to come within striking distance. Placing 300 larvae in a 400 litre tank makes it difficult to provide proper densities of live feeds/dry feeds with out polluting the water. Opportunistic contact between food and larva diminishes drastically when few are reared in a large tank. Forcing larva to swim considerable distances to seek food tends to drain their potential power supply quicker than it is replenished and results in eventual death or slow growth. Concentrating early stage larvae and food supplies minimizes production cost to provide sufficient live food organisms per unit area. By assuring more adequate food particles per unit area, energy expenditure of larvae to find food is minimized. This ensures faster growth rates and higher survivals.

Clown fish larvae are highly sensitive to light. High light intensities or sudden lighting induces stress to the larvae. The light intensity just sufficient to see the live feeds is preferred. An 8 watt CFL lamp hung at the middle of the tank gives adequate lighting. It is advantageous to have 24 hours light period to the larvae till 5 day. A lamp above the tank helps in concentrating larvae and rotifers to a particular place which helps in efficient feeding. Healthy clown fish larvae are clear yellowish in colour, with dark pigmentation around the stomach and eye region. Weak larvae remain dark along the entire body. Body shape of well fed larvae is oval or round. Larval mortalities are more common at hatch, day 2 after hatch when the yolk sac in almost gone, day 7-9 at metamorphosis and around day12. The most significant loses are on day 2 and 8. For the first 9-15 days, they are basically pelagic, scatter throughout the column. Early juvenile colouration is first detected by the development of the pale translucent white head bar. This occurs around day 7 to as late as day 15. Mortalities at metamorphosis are directly related to feed and water quality.

The clown fish reach the juvenile stage, which can be transferred to grow out tanks around 12-14 mm size when they are about 25-30 days old. Generally it takes a total of 4 months to rear to a marketable size of around 25-30 mm. Juvenile growth and development are strongly influenced by water quality, food quality and the amount of food fed. It is during juvenile growout that filtration capabilities become vital. A conventional submerged undergravel filter is not advisable in growout tanks. It is advantageous to use a bare tank with a single large airlift sponge filter or a bucket filter. The juveniles can be transferred very carefully with fine meshed net, to growout tanks. Clown fishes are territorial at very early age. The territorial problems can be prevented by providing a highly diverse habitat, increasing the tank volume so that each fish has several litres of water and crowding them so that there is no territory to defend. The last method is preferred due to practical reasons of cost, space requirements and maintenance troubles. Juveniles should be fed a minimum of 3 times a day to obtain maximum growth. Uneaten food and fecal matter should be removed at morning hours by siphoning.

Harvesting large individuals from a single tank can go on for several weeks and then it advisable to cull remaining fish and condense them for a final growout. Intense colouration is primarily developed through food. Fish grown in a very large tank (low density) have better colour than those in crowded conditions. Fish grown in dark backgrounds develop dark colouration while those in light coloured have light pale colouration. It is well known that astaxantin is the key pigment in clown fish. Products containing significant amounts of astaxanthin are most effective in enhancing pigmentation in clown fish. Frozen, freeze dried planktonic krill, lobster eggs, freshwater crayfish eggs and Macrobrachium eggs are also good sources of astaxanthin. Manipulation of diet, exterior environments, lights, water maintaining healthy and unstressed fish can all contribute to colouration.

Clown fish juveniles are most susceptible to disease and health problems. Constant vigil helps to diagnose the problem early. Sudden mortalities are usually due to deteriorated water quality conditions. Parasitic diseases cause slow constant death patterns whereas bacteria infestations lead to sudden large scale mortalities. Disease mainly occurs as a result of prolonged stress and weakening of the reared animals. Best way to control diseases is to prevent them by keeping constant attention on water quality and quantity of food provided to the fish.

References

Dawes, J., 1999. International experience in ornamental marine species management, Part 2: Some resource management strategies. *OFI Journal,* 27: 10–12.

Hoff, F.H.Jr., 1996. *Conditioning, Spawning and Rearing of Fish with Emphasis on Marine Clown Fish.* Aquaculture Consultants Inc., 33418 old Saint Joe Rd., Dade City, FL 33525.

Moe, M.A., 1999. *Marine Ornamental Aquaculture.* First International Conference of Marine Ornamentals, Hawaii.

Wood, E.M., 2001. *Collection of Coral Reef Fish for Aquaria: Global Trade, Conservation Issues and Management Strategies.* Marine Conservation Society, UK, 80 pp.

Chapter 13

Hatchery Production Technology for Clown Fish in Tamil Nadu

T.T. Ajith Kumar and T. Balasubramanian

Centre of Advanced Study in Marine Biology

Annamalai University

Parangipettai – 608 502, Tamil Nadu

In olden days, keeping marine ornamental fish was an impractical proposition due to its complexity in maintenance. But the recent developments in aquarium technology and better understanding of the biology and ecology of aquarium inhabitants made these organisms within the reach of aquarists. The coral reef provides a variety of ecological niches which are the abode of extremely rich and varied animal communities consisting of a great diversity of species. More than fifty fish families consisting of nearly 175 genera and about 400 species of ornamental fishes are distributed in the Indian seas. Unlike freshwater ornamental fish, which are mostly hatchery produced, the marine ornamental fish trade is sustained almost entirely by collection from the coral reef habitats. About 90 per cent of the freshwater fishes are farmed while 10 per cent are collected from the wild. But in the case of marine ornamental fishes, 92 per cent are collected from the wild and 8 per cent are captive bred.

Marine Ornamental Fish Trade

The marine ornamental fish trade has been expanding in recent years and has grown into a multimillion dollar enterprise mainly due to the emergence of modern aquarium gadget like synthetic sea water, canister filter, denitrifier, ozonizer and protein skimmer and technologies for setting and maintenance of miniature reef aquaria. The worldwide market of marine ornamental fishes has shown a steady increase over the past few years and the annual trade varies between US $ 2 and 7 billion. Over 1,800 species of fish were traded globally of which majority are originates from marine environment including crustaceans and invertebrates. Out of the traded species, only 25 per cent are bred under captivity and out of that only 21 species are commercially produced. The most commonly traded family of marine fish was Pomacentridae, which accounted for 43 per cent of all fish traded. Based on the Global Marine Aquarium Database (GMAD), the annual global trade is between 20 and 24 million numbers of ornamental fish, 11-12 million numbers of corals and 9 -10 million of other ornamental invertebrates.

Need for Captive Culture

In recent years, the surge in the trade of tropical ornamental fishes has increased considerably and at the same time indiscriminate exploitation has also led to negative repercussions on coral reef ecosystem. The eventual respond to a long term sustainable trade of marine ornamentals can be achieved only through the culture technologies. Hatchery production of marine ornamental is an environmental sound alternative to support this industry so that harvesting them from their natural habitat can be minimized and it is also a way to marine biodiversity conservation. In this scenario, the only possible alternative is the captive propagation of target species which ultimately resulting in decreased dependence on wild caught specimens which would also help to safeguard coral reef and develop a new source of organisms for the aquarium trade.

Advantages of Clown Fishes for Mass Production

Among the marine ornamentals, apt species for captive breeding for the aquarium trade is the clown fishes, due to their attractive colouration, peaceful nature, hardiness, proclivity to live in association with sea anemones. The main advantages of them are the compatible size, high market demand, spawn routinely and have large sized larvae. They belongs to the family Pomacentridae, is one of the largest groups of reef fishes, inhabiting tropical and sub-tropical seas and members of this family incorporate 29 genera and 350 species under four sub families. Totally 29 species of clown fishes are identified under two genera *Amphiprion* and *Premnas.* They are morphologically and taxonomically well distinguished from the damsel fishes due to their dependency on sea anemone for their protection.

Reproductive Biology of Clown Fish

Age of the fish is most important factor determining sexual maturity and the adult clown fishes are usually 9 - 18 months old. While selecting or establishing a pair, it is not advisable to purchase full grown adult fish and best to buy sub-adults. Even though they are younger than adult pairs, they will take about 3 - 6 months for initial spawning. A distinct advantage when pairing clown fish is their ability to change sex. The best and easiest approach in procuring fish for breeding is to select 3 or 4 fishes of equal size. Since, the sex reversal is prevalent in clown fish, they simply decide which will become the male and female. Under hatchery conditions, these fishes have to maintain under ideal environmental parameters and nutritious food with host sea anemones. The fishes are sensitive to changes within their environment, which includes water quality, photoperiod, light intensity, ammonia, etc.

Acclimatization and Pair Formation

Subsequent to selection, the fishes will put in quarantine and latter to the acclimatization tank along with sea anemones of 1 ton water holding capacity. After 5 - 7 weeks of rearing, one pair will grew ahead of others and become the spawning pair in which one fish make dominance over another one, which is female. Followed by this, other pair will form and the same can be transfered to 750 liter rectangular FRP tanks (spawning tank) containing 600 liter of water. The tanks will be provided with individual underwater biological filter and an artificial light. Each brooder tank may be provided with white coloured tiles, dead coral pieces and live rocks as substratum for egg deposition and imitates the natural environment.

Figure 13.1: Hatchery Technology for *Amphiprion clarkii* (Clown FIsh)

A brood pair of clown fish

Spawning activity

Females laying eggs

Clown fish larvae

Juveniles of clown fish

Marketable size clown fish

Figure 13.2: Hatchery Facility for Clown Fish Seed Production

Feeding

The fishes and anemones should be fed two times per day with boiled meat of shrimp, oyster, green mussel and clams at the rate of 4 per cent of their body weight. The broodstock need to be fed with fish egg mass and can also be provided formulated feeds enriched with vitamins and minerals.

Spawning and Incubation

Under the controlled conditioning regimes and proper feeding, each pair will start breeding within a period of 3 - 4 months of rearing in spawning tank. The broodstock diets are virtually the main key to successful spawning. The pair exhibit typical courtship behaviour of chasing each other and cleaning the substratum as indication for egg laying. Few days before spawning, the male selects a suitable site near to sea anemone for laying the egg and clear algal and debris with its mouth. On the day of spawning both the parents spent considerable time for cleaning the site which indicated that spawning may occur within few hours. Usually the spawning occurs at morning hours (9 to 11 am) and lasting for forty five minutes to one and a half hour. Female fish first lay capsule shaped eggs on the cleaned substratum in nearly rounded or oval patch. The male subsequently fertilizes the eggs. The clown fish have attached eggs and are known to spawn on rough surfaced substrata near to the host sea anemone. Hence, it is very essential to provide suitable substratum preferably tiles or earthen pots or shells of edible oyster or PVC pipes for the egg deposition. The colour of newly laid eggs varies from bright orange to yellowish. from 3rd day onwards, they start blackening and from 7 to 8th day they become silvery in colour due to the glowing eyes of the developing larvae inside the egg capsule. The incubation period ranges to 7-8 days depending on the surrounding environment. Generally the egg size of clown fishes ranges between 2.0 to 3.0 mm in length with a width of 0.8 to 1.5 mm, depends the species and size of the parents. An estimated average monthly fecundity is 300 to 1500 eggs per spawning.

Parental Care

As the parental care is inevitable for hatching out of the larvae, both the parent should be allowed to remain in the spawning tank itself till hatching. During incubation period, both the parents, especially male, carefully look after the eggs during day time and it involved two basic activities such as fanning the eggs by fluttering the pectoral fins and mouthing to remove the dead or weakened eggs and dust particles.

Larval Rearing

On the expected day of hatching, the brooder tanks are provided with complete darkness for accelerating the hatching. The newly hatched free swimming larvae are measuring 3 - 4 mm length and have 0.3 - 0.4 mm mouth size. They possess a transparent body, large eyes and small yolk sac. The newly hatched larvae individual initially sinks to the tank bottom, but quickly swims to the upper surface of the water column and this process called phototaxis. The larval stage of clown fish ends when its young ones settle to the bottom of the tank approximately 20 - 25 days after hatching.

After hatched out, the floating larvae must be transfered to small larval rearing tanks (100 litre capacity, FRP) with the density of 3 - 5 larvae/liter. The tanks should be provided with sufficient mild aeration, 14 hours light and algal enriched rotifer, *Brachionus plicatilis* at the concentration of 6 - 8 nos/ml. During larval stage, 10 per cent water exchange is done once in a day with bottom cleaning without disturbing the larvae. The metamorphosing period is in between 15-17 days and once it is started, all individuals metamorphosed with in 1 or 3 days. On the 10th day, the larvae can start to feed on newly hatched *Artemia nauplii.* The larvae are pelagic in the first two weeks and they move towards the bottom as they approach metamorphosis. From larval to sexual maturity and spawning the clown fish takes 9–15 months, depending on the species.

Live Feeds

The major constraint in rearing of any marine ornamental fish larvae is the heavy mortalities at different stages of development due to various factors. The size and nutritional quality of live feed are the two major factors affecting the survival of larvae. Nutritional quality of live feed depends on its unsaturated fatty acids contents. Fish larvae fed with live zooplankton indicated higher survival rate and increase in length and body weight than the artificial diets.

The rotifer, *Brachionus plicatilis* (SS) with 70-239 μm size can be used as initial feed to the larvae. Rotifers can be raised with the help of micro algae, *Chlorella* and *Nanochloropsis* spp. The stock culture of micro algae is able to maintain by using Conway medium and for the mass culture, agriculture fertilizers such as ammonium sulphate, super phosphate and urea at the ratio of 10:2:2.

Conclusion

The hatchery production and culture of marine ornamental fish can prove to be more economically feasible than that of marine food fish culture. This is because, even though the market for ornamental fish is much smaller than that of food fish and the price per unit is far higher in aquarium fish. Hence in future, hatchery reared fish will become a significant part of marine ornamental fish trade, because, this sector offers good opportunity for rural and urban households to augment income and link them to the international trade.

Acknowledgements

The authors express their deep sense of gratitude to the Ministry of Environment and Forests, Govt. of India for the financial assistance and the authorities of Annamalai University for facilities provided.

References

Allen G.R., Drew, J.A. and Kaufman, L., 2008. *Amphiprion barberi*: A new species of anemonefish (Pomacentridae) from Fiji, Tonga and Samoa. *Aqua,* 14: 105–114.

Allen, G.R., 1991. *Damsel Fishes of the World.* Mergus Publishers, Mello, Germany, 271 pp.

Aspari Rachman, 2010. Status of breeding, farming and trade of marine ornamental fishes in Indonesia. *Souvenir, Ornamentals Kerala–2010,* Dept. of Fisheries, Govt. of Kerala, p. 7–10.

Dey, V.K., 2010. Ornamental fish trade: Recent trends in Asia. *Souvenir, Ornamentals Kerala–2010,* Dept. of Fisheries, Govt. of Kerala, p. 39–45.

Fautin, D.G. and Allen, G.R., 1992. *Field Guide to Anemone Fishes and their Host Sea Anemones.* Western Australian Museum, Perth. pp. 160.

Gopakumar, G., 2006. Culture of marine ornamental fishes with reference to production systems, feeding and nutrition. *Souvenir, Ornamentals Kerala–2006,* (Eds.) Kurup *et al.* Dept. of Fisheries, Govt. of Kerala, p. 61–70.

Madhusoodana Kurup, B. and Antony, P.J., 2010. Indigenous ornamental fish germ plasm inventory of India with reference to the need for a paradigm shift of the industry from wild caught to farmed stock. *Souvenir, Ornamentals Kerala–2010,* (Eds.) Kurup *et al.* Dept. of Fisheries, Govt. of Kerala, p. 47–59.

Madhu, K., Madhu, Rema and Gopakumar, G., 2010. Breeding technology developed in marine ornamental fishes under captivity in India. *Souvenir, Ornamentals Kerala–2010,* (Eds.) Kurup *et al.* Dept. of Fisheries, Govt. of Kerala, p. 103– 112.

Chapter 14

Induced Breeding Protocols for Ornamental Brackishwater Fishes

M. Kailasam, A.R.T. Arasu, J.K. Sundaray, G. Biswas, Premkumar, R.Subburaj and K. Thiagarajan

Central Institute of Brackishwater Aquaculture No. 75, Santhome High 7Road, R.A. Puram, Chennai – 600 028

Introduction

Ornamental fishery resources face a range of challenges because of over exploitation from the wild and irrational fishing methods. It is high time that the strategy have to be worked out comprehensively for conservation and sustainable use. It is important to take the remedial measures for the problems caused by habitat loss and degradation, harmful fishing practices and changes in international trade patterns and concerns about the introduction of the exotic species. Aquarium keeping is one of the most practiced hobbies among the people of all age groups irrespective of the region and considered as an important commercial activity in the world. Having natural multicolour pattern, variation in shapes, dots, lines, spots, stripes, patches, slow swimming behavior and adaptability for living in small tanks, the aquarium fish are the most attractive pets for the people of all the age group. It is also believed that the hobby of aquarium keeping can provide peace of mind and recreation for the people who are very busy with their daily works by diverting them from their working stress. The growing awareness of environmental and nature issues has added to the attraction of fish keeping among public. Thus, there is great potential to popularize this hobby and to expand the ornamental aquarium fish trade both in domestic and international markets. In general, ornamental aquarium fish are obtained from freshwater, brackishwater and marine sources. The freshwater aquarium fishes have been well documented. Important attractive groups are produced in large quantities, with biotechnological applications. The marine coral fishes have also been contributed significantly in the ornamental fish trade. The brackishwater ornamental fish species, which can survive very well either in freshwater or marine/brackishwater, can also contribute significantly to the ornamental fish trade for the people living in coastal or inland regions. Identification, characterization and assessment of the suitability of the brackishwater ornamental fishes are the key factors in order to increase the aquarium fish production.

Commercially Important Brackishwater Ornamental Fishes

Although, there are several species of brackishwater ornamental fishes available in the brackishwater environment, only few are considered as economically important species because of market demand. Some of the commercially important candidate species are given below.

Sl.No.	*Species Name*	*Common Name*
1.	*Scatophagus argus*	Scat, Argus fish, Spotted Scat, Green Scat, Red Scat
2.	*Therapon jarbua*	Target fish
3.	*Etroplus maculatus*	Orange Chromide
4.	*Etroplus suratensis*	Green Chromide/Pearl Spot
5.	*Monodactylus argenteus*	Moon fish, Silver Angel, Monos
6.	*Takifugus ocellatus*	Fugu Puffer fish
7.	*Toxotes chatareus*	Large Scale Archer fish
8.	*Toxotes jaculatrix*	Banded Archer fish
9.	*Tetradon* sp.	Green spotted Puffer fish

Brackishwater ornamental fish can contribute significant share in the trade if the resource is utilized well planned planner. At present, the aquarium trade relies on capture for 98 per cent of its production versus 2 per cent culture whereas in the freshwater aquarium trade makes up 90 per cent of the total aquarium trade and 98 per cent of that is cultured. It is very important to develop suitable breeding technologies of these fish species in order to produce consistently to meet the growing demand for ornamental fishes. Artificial propogation of fishes is being practices for several food species to produce the seed for culture. Similarly, brackishwater ornamental fishes also can be bred either by hormonal manipulation or by environmental manipulation.

Induced Breeding Techniques

The reproductive activities are integrated with seasonal–environmental cycles in most of the aquatic species. The environmental exogenous factors such as temperature, photoperiod and rainfall, along with the endogenous physiological cycles send signals to the neuroendocrine system; which in turn regulate the pituitary–gonadal functions. If fishes kept under captivity in confined tanks for breeding, the change of environment from natural sea to a hatchery tank/pond can have a significant effect on hormonal regulation of gonadal function. The most common problem in the captive land based broodstock is the inhibition of final stages of oocyte and sperm development though early maturation takes place. The release of these gametes is referred to as spawning. Individual variation in terms of maturation among the brood stock fishes under captivity is observed. Some of the fishes attain mature gravid condition without any hormonal intervention and some did not.

Reproductive system is very much complicated in hermaphrodite fishes since they go through different phases of hormone secretion which is responsible for gonadal development. Maturation process can be induced/accelerated either by simulating the environmental conditions prevailing in sea or through the administration of the hormones responsible for maturation and spawning. However, simulating sea environment in many cases is difficult.

Hormonal Manipulation

The knowledge on the endocrine mechanism and the reproductive hormones is essential for inducing the maturation

Spotted Scat Brooders

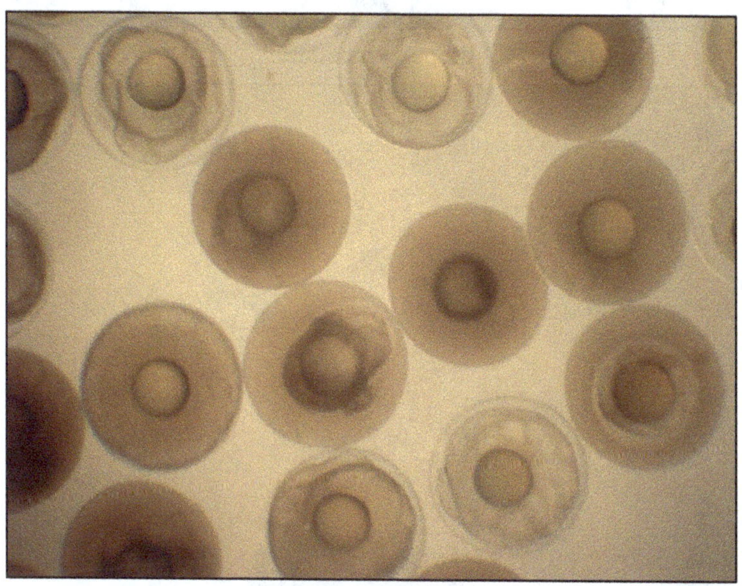

Spawned Eggs

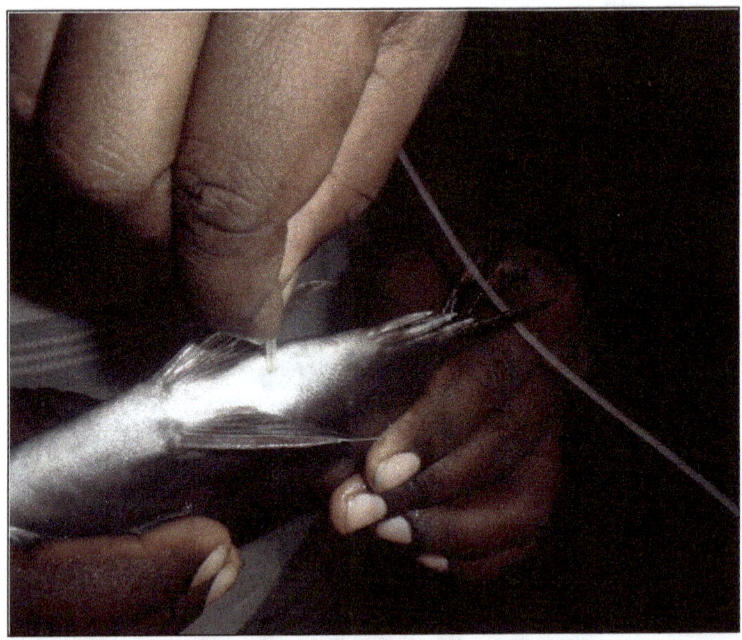

Biopsy Procedure

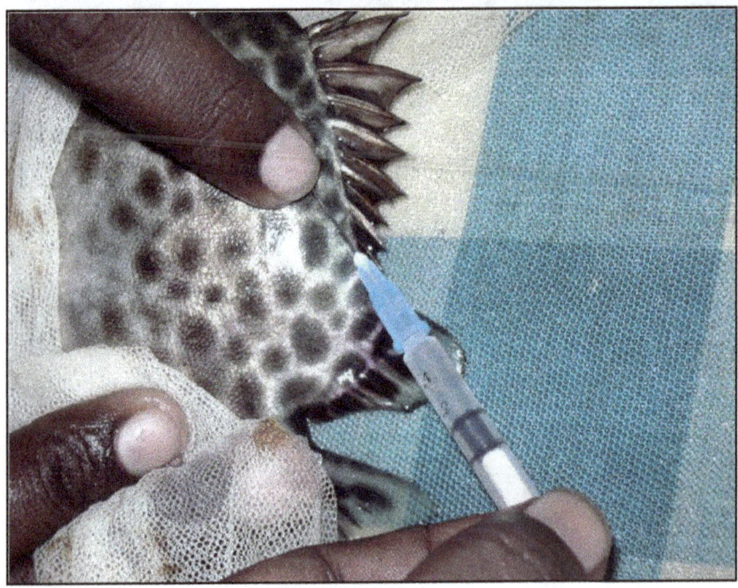

Hormone Injection

Tagging

process under captivity. Maturation requires stimulation by chronic and slow increase in hormone levels. This sustained slow release hormone delivery can be enhanced through external sources in two ways:

Hormone through Feed

Although most common and easy method of administration is through feed; it has certain limitations.

1. Only hormones that are not susceptible to enzymatic degradation in the digestive tract can be used.
2. There will be loss or hormone while this feed is in water.
3. No guarantee for absorption of hormone across the wall of intestine.
4. No control over the dosage administered as it depends upon the feeding rate of the individual fish.
5. Due to the above mentioned losses; excessive hormone has to be given.
6. Fish will revert back if hormone is not supplied continuously.

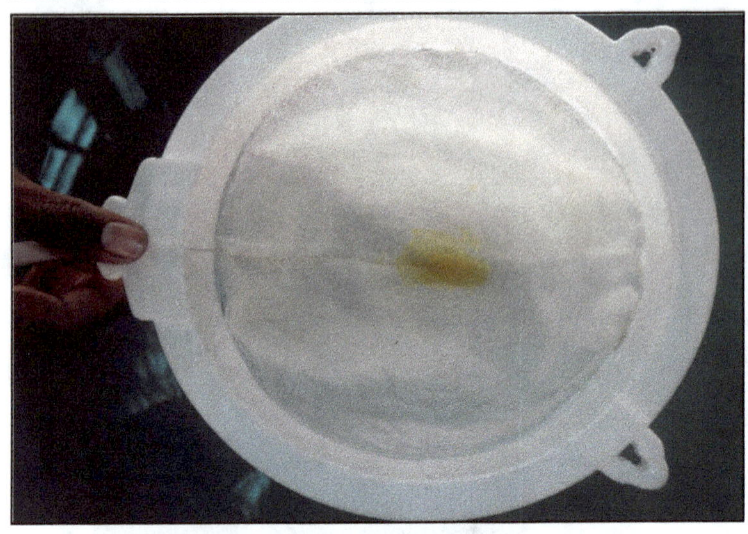

Spawned Eggs

Recirculation Facility for Captive Broodstock Development of *Scatophagus argus*

Hormone through Implantation

Search for chronic hormone delivery mechanism has produced a variety of pellets and capsules which help in slow release of hormones, when implanted into the musculature or abdominal cavity. These implants would assist in overcoming the short bioactive like of the hormones and induce constant elevation of gonadotropin secretion accelerating the final maturation of gonads. This technique has both advantages and disadvantages.

1. High risk is involved for life of the fish since implantation has to be done by making incision.
2. Sustained release of hormone makes the fish to respond continuously.
3. Implantation has to be repeated whenever necessary.

Therefore it is important to study the reproductive stages of these fishes before application of hormones.

Selection of Breeders

Breeders have to be selected from the captive broodstock before the onset of the breeding season, so that they can be conditioned to the environmental and diet controls. Breeders selected should be active, fins and scales should be complete, free from diseases, parasites and wounds. Males and females are selected based on their respective gonadal maturation stage. The maturity assessment of the fishes is done by ovarian biopsy as given below:

1. The fishes in the broodstock tank are secured individually for observation. In case the fishes are agitated too much or if wild broodstock fishes are procured, they can be transferred to anaesthesia tank containing sea water with 150 ppm of anaesthetic (MS–222 or 2-phenoxy ethanol). After turning over the fish; abdomen is gently massaged in head to tail direction. If milt extrudes out of the genital pore, it is a ripe male.
2. If milt is not there, a polyethylene canula of 1.2mm diameter is inserted into the genital opening of the fish upto 5 cm. The other end is aspirated gently by mouth while withdrawing the canula slowly.
3. The contents are collected in a watch glass and observed microscopically. Measuring the ova diameter using an ocular micrometer assesses the stage of maturing.

4. Female with <0.4 mm average egg diameter can be chosen for pellet implantation and the males without milt are also chosen for pellet implantation. Females tertiary yolk globule stage or a ova diameter of 0.45–0.50 mm are chosen for breeding purpose.

Pellet Preparation

The sexual maturation can be accelerated by implanting Leuteinizing Hormone Releasing Hormone–analogue (LHRH–a) pellet. The hormone is incorporated in a matrix of cholesterol powder and made into a pellet of required diameter for implantation. Before the pellet preparation, fish is weighed and the dose is calculated accordingly. Dose of LHRH-a to be implanted is calculated @ 100µg/kg body weight of the fish. Weigh the cholesterol powder @ 0.25g per can be taken in a clean, dry petridish.

1. Dissolve the hormone in 0.2ml of 80 per cent ethanol. Draw the solution with a syringe.
2. Mix the cholesterol powder with dissolved hormone with a spatula. Equal quantity of pinch of cellulose or gum acacia is added as a binder.
3. Weigh the total amount of powdered mixture containing the hormone required for each fish.
4. Compact the powdered mixture into cylindrical pellets by using pellet maker of required diameter, usually 0.2-0.3mm.
5. Dry the pellet in hot air over at 37°C for 2-3 hours or at room temperature for a day and stored in capped vials.

LHRH-a or MT (17-µ methyl testosterone) pellets can also be prepared in the above mentioned manner and used for the induction of maturation in males. This case is applicable only if adequate males are not available. By treating the females through methyl testosterone it can be converted into males.

Pellet Implantation

Intramuscular Implantation

1. The fish to be implanted is isolated and transferred to an anaesthesia tank, if agitated too much.
2. Turn the fish laterally and removed one or two scales with a forceps at a point 2-3 cm below the dorsal fin.

3. A short incision of around 1 cm width and 1 cm depth is made by using a surgical blade. Incision is made perpendicular to the normal spread of the muscle fiber.
4. The pellet is inserted into the musculature with the help of a forceps through the incision made and twisted at 90°C angle to get itself embedded in the musculature.
5. After withdrawing the forceps, the incision is sealed with pure Vaseline and the fish is given antibiotic treatment to prevent infection.
6. The fish is released in the broodstock tank after properly tagging for identification.

Hormone pellet implanted fishes are examined fortnightly to keep track of the gonadal development. Normally within 1-2 months after implantation, maturing fishes attain gravid condition. If needed, implantation can be repeated fortnightly/monthly once until the onset of breeding season.

The environment conditions prevailing in the sea may be essential for activating the hormones responsible for reproduction. Matured fish can be made to spawn under captive condition, if the sea-conditions can be simulated. But simulating such conditions in hatcheries has limitations and it is not possible in all the places. However by administering extraneous hormones responsible for ovulation and spawning will be helpful to induce spawning. Many hormones like, LHRHa, HCG, ovaprim, ovatide and carp pituitary extracts can be used.

Induced Breeding of Spotted Scat *Scatophagus argus*

Selection of Spawners

The spotted scat *Scatophagus argus* (L.) 1766 (Perciformes: Scatophagidae) is considered as an important fish in ornamental fish trade owing to its unique colour pattern having greenish brown to silvery, quadrangular body shape with many brown to red brown spots and dark vertical bars. It is also a preferred food fish in Southeast Asian countries especially in the Philippines, where it is considered as an expensive delicacy. Development of breeding technology for spotted scat would help to promote the ornamental fish trade in coastal region.

Broodstock fishes have to be maintained in the recirculation facility with proper feeding and water quality management in order to achieve the maturation. The brood fishes held under captivity in the recirculation tank were subjected for assessing the maturity. Females with size of above 150g can be subjected for ovarian biopsy. The females should have the oocyte diameter of above 450µm, which can be used for induction of spawning. Males with the size of above 80gm can be noticed with the oozing of milt can be used for the spawning purpose.

Male and female ration @ 2:1 prior to induction of spawning can be kept together in the spawning tank (400 L capciaty circular tank) in order condition them. Mud pots or PVC pipes can be used as hide outs by keeping them in the spawning tank. The tank can be set with the flow through facility and mild aeration can be provided. Water quality parameters such as temperature, salinity, pH, dissolved oxygen and ammonia recorded from the broodstock tank can be maintained within the range of 28-31°C, 20-25 ppt, 7.78-8.01, 4.40-5.10 ppm and 0.06-0.12 ppm respectively.

Human Chorionic Gonadotrophin (HCG) and Leutinizing Hormone and Releasing Hormone (LHRHa) can be used for induction of spawning. Females can be administered with priming dose of HCG @ 750–1000 IU/kg BW and after 24 hrs LHRHa can be administered @ 100 µg/kg BW. Pairing behavior and courtship of male following the female was observed and often as one of the males made aggressive attack on the other male and this antagonistic behavior seen initially subsided. Later, other male positioned opposite to female. Courtship concluded while the male biting and holding up the upper lip of the female by the males with greater frequency till spawning. Female spawned spontaneously after 48 hours of LHRHa hormone injection. A total of 1.73 lakh eggs could be collected from the spawning. The ovulated and spawned eggs settled at the bottom and the eggs appeared opaque. The mean diameter of the spawned eggs was 770±8µm.However, the spawned eggs were unfertilized. Further trials to fertilize the eggs are under progress.

Conclusion

Development of breeding technology for brackishwater ornamental fished is gaining significant important in the recent years because these fishes can be maintained easily both in freshwater and seawater conditions. Species such as Moon fish *Monodactylus argenteus,* spotted scat *Scatophagus argus,* puffer fishes etc., are in great demand for aquarium purpose. If the breeding technology is developed for these species, the technology can be imparted to the farmers and self help groups and the beneficiaries can be expanded. Brackishwater ornamental fish trade had great potential in India and in order to meet the demand for aquarium purpose, the breeding technology have to be developed essentially, which would also reduce the pressure from exploiting the natural resources.

Part V

Technology Dissemination and Trade

Chapter 15

The Role of Fisheries Department in Promoting Aquariculture Sector

R. Thillai Govindan
Joint Director of Fisheries (Inland Fisheries)
Government of Tamil Nadu
Teynampet, Chennai – 600 006

Ornamental fish keeping is one of the most popular hobbies in the world today. The growing interest in aquarium fishes has resulted in steady increase in aquarium fish trade globally. The trade with a turnover of US $ 5 Billion and an annual growth rate of 8 per cent offers a lot of scope for development.

The top exporting country being Singapore followed by Hong Kong, Malaysia, Thailand, Philippines, Sri Lanka, Taiwan, Indonesia and India. The largest importer of Ornamental fish is the USA followed by Europe and Japan. The emerging markets are China and South Africa. Over US $ 500 million worth of ornamental fish are imported into the USA each year.

Ornamental fish keeping and its propagation has been an interesting activity, which provides not only aesthetic pleasure, but also financial opportunities. About 600 ornamental fish species have been reported worldwide from various aquatic environments. Indian waters possess a rich diversity of ornamental fish, with over 100 indigenous varieties, in addition to a similar number of exotic species that are bred in captivity.

India's share in ornamental fish trade is estimated to be Rs 158.23 lakh which is only 0.008 per cent of the global trade. The major part of the export trade is based on wild collection. There is very good domestic market too, which is mainly based on domestically bred exotic species. The overall domestic trade in this field cross 10 crores and is growing at the rate of 20 per cent annually.

The earning potential of this sector has hardly been understood and the same is not being exploited in a technology driven manner. Considering the relatively simple technique involved, this activity has the potential to create substantial job opportunities, besides helping export earnings.

Starting an Ornamental Fish Farm

For a beginner, one should start with species that are inexpensive and grown easily, particularly those that are born live. After learning the biology, feeding pattern and optimum condition for raising live-bearing fish, one should try rearing egg-layers. It is also wise to concentrate on growing a single species while learning. To begin with trial may be made for raising guppies.

One can pursue ornamental fish farming by raising broodstock for sale to fish farmers, rearing fish from broodstock or by doing both. So that one could be master in all the aspects of growing fish. It is suggested to grow the young ones until maturity, which could take around four to six months of time.

Breeding of Ornamental Fishes

Ninety five per cent of countries ornamental fish export is currently based on wild collection. Majority of the indigenous ornamental fish trade in India is from the North Eastern states and the rest is from Southern states which are the hot spots of fish bio diversity in India. This capture based export is not sustainable and it is a matter of concern for the industry. In order to sustain the growth it is absolutely necessary to shift the focus from capture to culture based development.

Moreover most of the fish species grown for their ornamental importance can be bred in India successfully. Organized trade in ornamental fish depends on assured and adequate supply of demand, which is possible only by mass breeding. A beginner should start the work on breeding of any live-bearer followed by goldfish or any other egg-layer for getting acquainted with the procedures on handling and maintenance of brood fish and young one. Good knowledge on the biology, feeding behavior and ambient condition of the fish are prerequisites for breeding. The method of breeding is based on the family characteristics of the fish.

Breeders especially egg layers should be discarded after few spawnings. Health care, water exchange is a must for maintaining water quality conducive for the fish health. Only healthy fish can withstand transportation and fetch good price.

Young fish are fed mainly with Infusoria, Artemia, Daphnia, Mosquito larvae, Tubifex and Blood worms. For rearing, formulated artificial or prepared feed can be used. At present no indigenous prepared feed for aquarium fish is available. The amount and type of food to be given depends on the size of the fry. Feeding is generally done twice in a day or according to requirement. For rearing from fry stage dry/prepared feed can be used. There is a good domestic market which is increasing. The export market for indigenously bred exotic species is also increasing.

Activities of the Department

Ornamental fish culture is encouraged by the department through the Schemes such as IAMWARM, NADP, WGDP, STATE SCHEME - subsidy assistance etc.,

IAMWARM (Irrigated Agriculture Modernisation and Water-Bodies Restoration and Management)

Under IAMWARM, the ornamental fish culture activity is being promoted as a backyard type unit in selected sub-basins as a model activity to improve the socio-economic status of the farmers. The proposed ornamental fish culture/rearing units shall be established in suitable sites owned by either Water Users Association or any progressive farmer. The beneficiary will be provided with necessary training, technical inputs and marketing linkage by the departmental staff. Economically important ornamental fish varieties will be reared from young to adult stage and marketed.

The proposed ornamental fish culture unit shall be established in 500 m^2 area. The unit should have a bore well, breeding/rearing cement cisterns, circular tanks, glass tanks and packing shed. Under the scheme 100 per cent grant is provided *i.e.,* Rs.2.00 lakhs per unit. So far, under this project in the last two phases 18 units were established in various sub basins of Tamil Nadu.

In the third Phase, 26 Nos. of ornamental fish culture units have been proposed at an estimated cost of Rs.52.00 lakhs (Rs.2.00 lakhs per unit) in the following sub-basins.

Sl.No.	*District*	*Sub-basin*	*No. of Units Proposed*
1.	Thiruvallur	Araniyar	4
2.	Thiruvallur	Kosasthalayar	4
3.	Kancheepuram	Ongur	4
4.	Thiruvannamalai	Kambainallur	1
5.	Virudhunagar	Kanal Odai	1
6.	Virudhunagar	Nagarier	2
7.	Virudhunagar	Sevalaperiar	2
8.	Dharmapuri	Deviyar	3
9.	Theni	Theniyar	3
10.	Sivagangai	Girdhamalnadhi	2
		Total	26

Western Ghat Development Programme (WGDP)

Ornamental fish culture is one of the major fish culture activity which can be easily taken up by rural folk even at their backyard. The Western Ghats has rich resource potential for wide variety of ornamental fishes. Under WGDP creation of infrastructure facilities through subsidy component and training in ornamental fish culture at Virudhunagar was provided to create awareness on ornamental fish culture as a rural self employment income generating activity, an amount of Rs.2.50 lakh was allotted by the District Collector, for which training was provided on ornamental fish culture at an amount of Rs.10,000/-. Later a group of farmers were selected from Rajapalayam and Srivilliputtur taluks in Virudhunagar district and hands-on practical training on freshwater ornamental fish culture were given to them. They were also exposed to an exposure visit to ornamental fish culture unit of TNFDC at Aliyar in Coimbatore District.

After training, six beneficiaries were motivated to take up ornamental fish culture and they constructed the farms on their own, as per the rough cost estimate provided by the Department. After construction, the farms were stocked with breeders and spawns of various ornamental fishes. On completion of all the facilities, the beneficiaries were provided subsidy to the tune of Rs.40,000/- each totaling an amount of Rs.2.40lakhs. Due to the hands–on training and the subsidy provisions, the beneficiaries are all now successful in culturing the ornamental fishes and they are now earning a minimum profit of Rs.6000-10,000 per month, which made the project a success. If the individuals belonging to the western ghat districts are interested in ornamental fish culture, they can approach the department so that according to their need schemes can be obtained and provided.

National Fisheries Development Board (NFDB)

The importance of ornamental fish sector in the world trade is that, it is a source of income, especially for the rural population in the country. Inspite of this, the existing ornamental fish market is highly unorganized in India and the absence of clear statistics about this industry, its related activities, species bred, reared, exported, imported and the people involved in this business.

In Tamil Nadu, Kolathur village near Chennai is famous for ornamental fish production with about 2000 families being involved in this activity with rural backyard type of hatchery and rearing. This has paved way for development of more rural backyard type of hatchery in other places such as Oothukottai and Usilampatti. Data on such farms is presently unavailable and quite essential for further development of this industry. Visualising this need, the department has approached the NFDB for a project on "Creation of Database on ornamental Fish Industry" at a cost of Rs.2.44 lakhs and the NFDB has sanctioned the same.

The department on completion of this project will bring out a create a database which would be useful for all the ornamental fish culturists, researchers, and public. This would pave way for further development in this industry as well as aid in planning future schemes for this sector.

For the year, 2010-11, the Department has proposed the following components, for the ornamental fish sector to help the ornamental fish culturists. Five batches of benifishery with 30 nos./ batch have been trained on ornamental fish rearing with 100 per cent grant of Rs. 1.5 Lakhs form NFDB.

Name of the Component	*Total units*	*Unit Cost (Lakhs) in Rs.*	*Subsidy/ Individual (Rs.)*	*Total Amount (Lakhs) in Rs.*
Establishment of backyard hatcheries for ornamental fish production	100	1/unit	25,000	25.00
to	10	8/unit	25,000	2.00
Aquarium fabrication	60	1/unit	25,000	15.00

The above scheme, will attract new ornamental fish culturists and will help in further expansion of this industry. Based on the success of the 2010–2011 projects, new avenues will be created for ornamental fish culturists by the Department.

Aquatic Animal Quarantine Laboratory

The Department has provided land at Padappai to Government of India for setting up an Aquatic Animal Quarantine Laboratory. This laboratory will be set up with International standards with a capital outlay of Rs.12.00 crores and facilities besides providing quarantine facilities for the fish importers and exporters. The site is located just within an hour's drive from Chennai Airport and will definitely help the ornamental fish industry.

Plan of Action for the State

1. The Department has already envisaged the growth of this industry and requested the men and women involved in this sector to farm societies so that Government benefits, credit assistance, Central and State sponsored subsidies can be availed.
2. There should be an integrated approach in this industry for long term survival *viz.*, there should be coordination in all the activities–brood stock development -> breeding -> nursery rearing -> feed production -> live feed and pellet feed production -> commercial production -> marketing -> accessoring market. All these components should be

interlinked so that the ornamental fish culturist reap better revenue.

3. Species diversification, breeding and production should be advocated for the survival of this industry on a long term basis.
4. Marketing is a main thrust area which secures the industry. An excellent example is the tie up arrangement being practiced in Gumidipoondi wherein, the buy-back arrangement is guaranteed for the growers and genuine prices is also being offered.
5. Customer education is lacking in this sector. The ornamental fish culturist should identify good customers, educate them based on their time, space and need this will ensure more number of aquarium keepers in the long run. In countries like Singapore and Malaysia, the aquarium hobbyists are thoroughly educated. Such system should be followed in our country also which in turn will fetch more revenue in the local market instead of relying on exports.
6. The research institutes should simplify the breeding methodology for native species and transfer the technology to the farmers.
7. Also the endangered ornamental species should be enlisted, breeding programme should be taken up and ranching the seeds in the natural environment should be made to improve the natural resources.
8. A coordinated, ecofriendly, sustainable system of approach among farming community, institutes and the department will definitely pave way for a sustainable "RAINBOW REVOLUTION" in the long run.

The Department has a farmer friendly approach to benefit every fish farmer. All assistance and further queries can be had at each district Assistant Director of Fisheries, Inland or Aquaculture office.

Chapter 16

Factors that Hamper the Ornamental Fish Production Sector in Tamil Nadu

N. Mini Sekharan, S. Vivekanandan, S. Asanarr, Soumya Subhra De
School of Industrial Fisheries
Cochin Univeristy of Science and Technology
Kochi – 682 016

Tamil Nadu is a highly productive state whose production is worth emulating in many aspects such as agriculture, fisheries and veterinary. The state possesses a remarkable history in ornamental fish breeding and can aptly be termed as the "Ornamental Fish Capital of India" due to its contribution to both domestic as well as export sector. The state is the prime contributor to the domestic ornamental fish marketing sector in India and caters ornamental fish right from the bred and butter varieties to very costly varieties to almost all states in India round the year. Tamil Nadu occupies the second position in ornamental fish export from India (Figure 16.1). This chapter is an outcome of the MPEDA/UNCTAD funded a project on "*Developing Strategies to Network the Ornamental Fish Breeders in India*

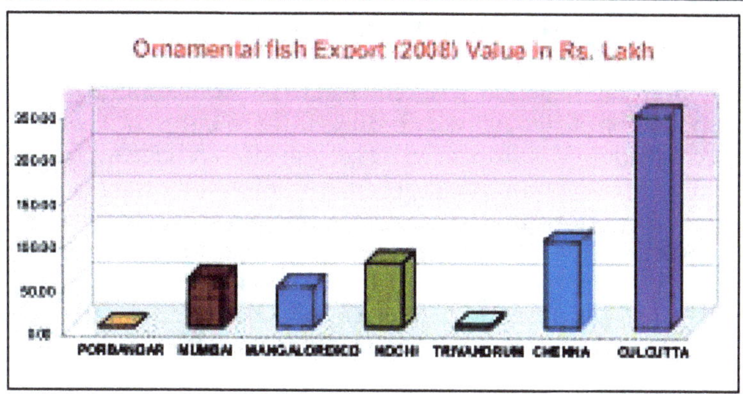

Figure 16.1: Ornamental Fish Export from Different Ports of India

for Enhancing Exports" carried out during 2007-2008 by the author who was the Principal Investigator of the project. The study focused on the objectives of surveying the ornamental fish breeders in five states of India (Kerala, Karnataka, Tamil Nadu, Maharashtra and West Bengal).

Background of the Study

Literature surveys revealed that papers on ornamental fisheries are specifically on fish related aspects and papers related to the ornamental fish sector related issues are very few. Papers could be traced on other sectors in Tamil Nadu *i.e.*, Cotton cultivators of Virudhunagar district, cutflower growers in Nilgiri district, perishable commodities with special reference to flowers in Srirangam Market, Thrichirappalli, vanilla cultivators in Coimbatore district, mulberry growers of Erode district. Therefore the need for a holistic study of the ornamental fish sector was perceived.

A number of issues holdback the breeding sector of India from enhancing the ornamental fish production. General problems mentioned as constraints for the sector is the lack of availability of brood stock, difficulty in marketing, lack of professional training in breeding, lack of training in nutrition and disease and health care of ornamental fishes. According to Kumar (2005) for the rapid growth of live ornamental fish industry of India, import of brood stock of different varieties is a pre requisite. As Government of India has placed ornamental fishes in the restricted item of import and enquires special import license for importing fishes and the obtaining of license and import is a herculean task for marketers. He put forward several recommendations which need to be seriously studied for simplification of quarantine procedures in the country. For a country which is endowed with extremes of geographical and climatic factors it would not be appropriate to generalize the issues that hamper any production sector. The study therefore, resorted to bring out specific issues of the ornamental fish breeders of 5 states in India. The present paper look into the issues specific to the state Tamil Nadu in ornamental fish production and each district of Tamil Nadu which had a high concentration of ornamental fish breeders.

Research Design

The study carried out survey by personal interview method which was in depth in nature. Questionnaire was developed as survey instrument for collecting primary data. It was structured in such a way that the primary objectives of the research could be addressed. 9 Survey Assistants and a Research Assistant (familiar with the local language of the chosen states) were selected and trained to carry out the study by personal interview method by visiting the breeding site of each breeder. For the survey in Tamil Nadu, 2 MSc Graduates from Annamalai University who were natives of Tamil Nadu and well versed in Tamil was seleced as they had to interact well with the breeders in different districts of the state. The data analysis was carried out using the SPSS.16 software.

Distribution of Ornamental Fish Breeders in India and Tamil Nadu

The ornamental fish breeders form the vital component of the ornamental fish industry. The project survey identified 1703 breeders in the five states of India. Of the total breeders surveyed, only 608 were full time breeders and 1059 part time breeders. Among the states surveyed, West Bengal dominated with 47.6 per cent of the total breeders/breeding units in India and was followed by Kerala (23.7 per cent), Tamil Nadu (17.5 per cent), Maharashtra (8 per cent) and Karnataka (3.3 per cent) (Figure 16.1). In Tamil Nadu, high percentage of the breeding units were located in Thiruvallur, Chennai followed by Dindigul and Madurai districts (Figure 16.2).

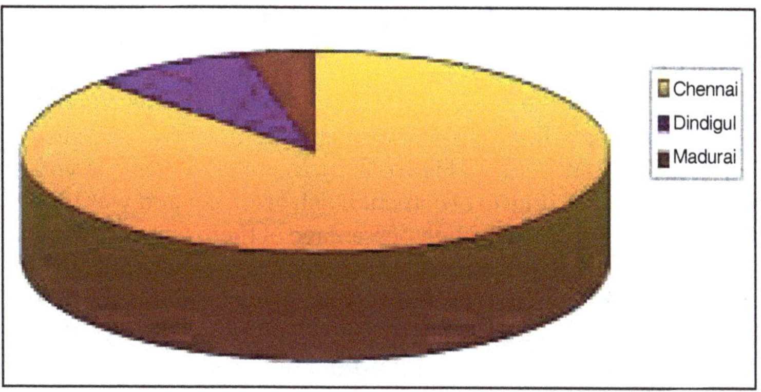

Figure 16.2: Distribution of Ornamental Fish Breeding Units in Tamil Nadu

Category of Ornamental Fish Breeders

A person carrying out breeding activity exclusively is generally termed as breeder. In India, any person involved in activities such as breeding, rearing/growing or combining breeding with marketing activities related to ornamental fish is also described as a "breeder". The breeder category in India is given listed below and the activity carried out by them is given in bracket,

- Breeders (only Breeding-B)
- Rearers/Growers (only Rearing-R)
- Breeder -Rearer (Breeding and Rearing-BR1)
- Breeder -Retailer (Breeding and Retailing- BR2)
- Breeder -Exporter (Breeding and Exporting-BE)
- Breeder-Rearer -Retailer (Breeding, Rearing and Retailing-BRR)
- Breeder-Retailer-Exporter (Breeding, Retailing and Exporting-BRE)
- Breeder-Rearer-Retailer-Wholesaler (Breeding, Rearing, Retailing and Wholesaling-BRRW)
- Breeder-Rearer-Wholesaler (Breeding, Rearing and Wholesaling- BRW)

Characteristics of Ornamental Fish Breeders of Tamil Nadu

On analysing the category of breeders in India it was noted that highest number of breeders, who carried out breeding exclusively, was found in Tamil Nadu (Figure 16.3). Majority of the breeders in the state were full time in ornamental fish breeding and was highly experienced (Table 16.1). High percentage of them was members of any group or association unlike the breeders in other states. They showed a high coordination among themselves and this sharing in fact might have played a key role in their gaining expertise as breeders. As per the study ornamental fish production from India was to the tune of 100 million fishes per year, of which more than 10 per cent of the production was the share from Tamil Nadu.

Barriers for Production Expansion

The total production from the surveyed farms in India is approximately 100 million fishes/year even though they have a capacity to produce double the number produced now. It was noted

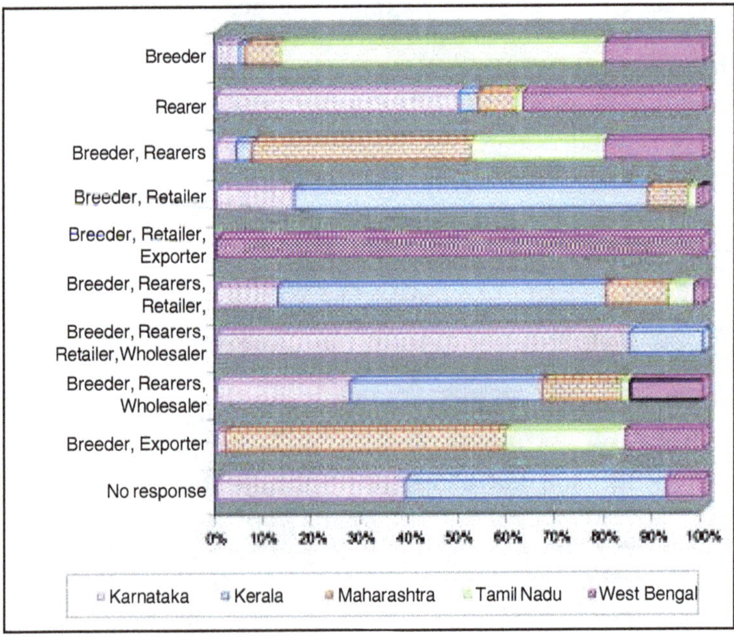

Figure 16.3: Classification of Breeders Based on their Activity

Table 16.1: Experience in Ornamental Fish Breeding

State	*Experience*						*Total*
	0-10 Yrs	*10-20 Yrs*	*20-30 Yrs*	*30-40Yrs*	*>40 Yrs*	*No Response*	
Karnataka	39 (69.6%)	11 (19.6%)	2 (3.6%)	0 (0%)	4 (7.1%)	0 (0%)	56 (100%)
Kerala	228 (56.6%)	13 (34.0%)	21 (5.2%)	(71.7%)	1 (0.2%)	9 (2.2%)	403 (100%)
Maharashtra	79 (58.1%)	47 (34.6%)	6 (4.4%)	1 (0.7%)	1 (0.7%)	2 (1.5%)	136 (100%)
Tamil Nadu	156 (52.3%)	111 (37.2%)	24 (8.1%)	2 (0.7%)	2 (0.7%)	3 (1%)	298 (100%)
West Bengal	630 (77.8%)	146 (18%)	17 (2.1%)	6 (0.7%)	3 (0.4%)	8 (1%)	810 (100%)
Total	1132 (66.5%)	452 (26.5%)	70 (4.1%)	16 (0.9%)	11 (0.6%)	22 (1.3%)	1703 (100%)

Source: SIF Primary Survey, 2008.

Table 16.2: Membership in Ornamental Fish Association/Group

Sl.No.	*State*	*Association*			*Total*
		Member	*Non-member*	*No Response*	
1.	Karnataka	33 (58.9%)	22 (39.3%)	1 (1.8%)	56
2.	Kerala	85 (21.1%)	280 (69.5%)	38 (9.4%)	403
3.	Maharashtra	22 16.2%	111 (81.6%)	3 (2.2%)	136
4.	Tamil Nadu	229 (76.8%)	69 (23.2%)	0 (0%)	298
5.	West Bengal	506 (62.5%)	296 (36.5%)	8 (1.0%)	810
	Total	875 (51.4%)	778 (45.7%)	50 (2.9%)	1703

Source: SIF Primary Survey, 2008.

that among the five states surveyed Tamil Nadu topped in ornamental fish production. The state has more potential for enhancing production but faced a number of constraints. Kulathur in Tamil Nadu is hailed nationally and internationally as a unique model in ornamental fisheries development and contribution towards livelihood and still they had a number of issues that troubled them. The pre survey revealed the main barriers of production expansion in ornamental fish farming in India as difficulty in getting finance from institutions, high electricity charges based on industrial rate, difficulty in getting brood stock of pure strains of ornamental fishes, lack of infrastructural facilities, lack of market information, lack of steady demand, difficulty in marketing, lack of technical advice or good training and lack of space.

The surveyed breeders were asked to rank the constraints as 1 for the most difficult constraint to 8 for the least difficult constraint. On analyzing the mean of ranks for barriers in production expansion, it was noted that in the case of total ornamental fish breeders of India, difficulty in obtaining finance formed the main constraint (2.59), followed by high electricity charges at industrial rate (4.41), lack of market information (4.65), lack of pure strains of fishes (4.68) and lack of technical advice on different aspects of ornamental fishes (5.12).

Ornamental fish breeders of Tamil Nadu felt that the high electricity charges to be the major constraint in production expansion (1.01) followed by lack of technical advice (2.0) and difficulty in obtaining finance (3.04).

1. The domestic charge for electricity was Rs 1.50 per unit and for commercial usage the charge is Rs. 7.50. For agricultural usage the charge was Rs 3 in urban area and free of cost in rural areas. The ornamental fish breeders were not included under the agricultural sector or the small scale industry and hence breeders had to pay electricity charges as per the commercial industrial charge of Rs.7.50/ Unit. The breeders in Tamil Nadu strive hard atleast to get it included under Small Scale Industries (SSI) but in vain. The reasons why ornamental fish breeders could not be included under SSI or as agricultural sector needs to be studied and resolved at the earliest so that their persisting issues of high electricity charge can be solved for enhancing production.

Table 16.3: Barriers in Production Expansion in Different States of India

Barriers	*India (Mean)*	*Karnataka (Mean)*	*Kerala (Mean)*	*Maharashtra (Mean)*	*Tamil Nadu (Mean)*	*W. Bengal (Mean)*
High electricity charges	4.41(II)	5.65	3.41(III)	5.59	1.01(I)	7.13
Difficulty in obtaining Finance	2.59(I)	4.47(III)	2.65(I)	4.57	3.04(III)	1.52(I)
Difficulty in obtaining Pure Strains	4.68(IV)	4.84	2.9(II)	4.08(II)	7.35	4.36(II)
Lack of Infrastructure	5.92(VIII)	5.35	6.21	4.32(III)	7.94	4.99
Lack of Market information	4.65(III)	4.56	5.01	5.59	4.13	4.5(III)
Limited demand	5.63(VII)	4.28(II)	5.53	5.26	5.11	6.27
Difficulty in Marketing	5.93(IX)	2.86(I)	5.85	5.61	8.77	4.6
Space Constraint	5.47(VI)	7.1	6.55	3.92(I)	6.03	4.67
Technical Advice	5.12(V)	4.77	6.31	5.85	2(II)	6.14

Source: SIF Primary Survey, 2008.

2. Lack of technical advice was a problem faced by the breeders in Tamil Nadu. In Tamil Nadu most of the breeders expressed the need for expertise in disease diagnosis and treatment and also live feed culture.
3. In the case of obtaining of finance, official paper work for obtaining finance and the need for security was a major issue faced by the breeders, followed by need of feasibility reports as a proof of profitability, the disinterested attitude of bank due to the perishable nature of fishes and high interest rate. It was also noted that, financial institutions considered ornamental fisheries to be a risky and unprofitable business. Absence of proper land document was a major issue faced by many breeders while trying to apply for subsidies and loans. Banks were hesitant in giving loans owing to many reasons such as the doubtfulness regarding the profitability, absence of documents, etc. and sanction very less amount as loans.
4. Some of the other issues faced by the breeders were regarding live feed collection. In areas near Kolathur live feed collection was mainly carried out from Velacheri and nearby areas and there were people who were dependant on live feed collection for livelihood and this traditional way of collecting live feed had its social and health issues. Live feed collection at early morning hours of about 3 am often led to accidents and drowning. The breeders of the state wished for training on live feed culture on a large scale.
5. Floods during rains affected many of the breeding areas of Tamil Nadu and it was difficult for the breeders to obtain any kind of flood relief funds.
6. Water is not of good quality in many places in Tamil Nadu especially Chennai due to pollutants (high pH and hardness).
7. Extension activities rarely reached the breeders as there was a lack of extension personnels for ornamental fisheries in the State Fisheries or Marine Products Export Development Authority. There is a need for permanent staff to make field visits to the ornamental fish breeders to advice on fish health, subsidies and carry out calamity

assessments. It was noted that very less percentage of the breeders in Tamil Nadu had attended formal trainings on ornamental fish breeding but they had better experience and expertise.

8. The breeding sector in the state is characterized by small scale units with low investment lesser than 2 lakh and small size holdings. A large number of them function as backyard breeding units with limited infrastructure. The infrastructural facilities of the farms have to be enhanced in leaps and bounds for quality production. The breeders were upset by the fact that the subsidy schemes of MPEDA were in favour of the new breeders and there was no amount earmarked for the empowerment and development of existing firms.
9. Many breeders do not attach much importance to the quality factors as anything produced is taken up by the domestic markets. They try to breed large number of species and varieties rather than concentrating on few varieties.
10. Difficulty in obtaining brood stock was a main problem for all breeders as the import procedures for ornamental fishes was quite tough and there were no research centres supplying good broodstock.

Conclusion

Tamil Nadu was the only state in India which has maximum number of specialized breeders focused on ornamental fish breeding as a full time activity. The state also topped among the other states in the quantity of ornamental fish produced. Inspite of all the factors that make Tamil Nadu the ornamental fish capital of India, a number of factors hamper the ornamental fish breeders of Tamil Nadu from increasing production of ornamental fishes. Including the sector in small scale sector and subsidising the electricity and water charges and enabling an ease of availability of loan is the prime requirements of the breeders of Tamil Nadu. Capacity building at all levels is crucial if the sector has to raise its production. The breeding sector has to grow in response to the demand supply deficit for ornamental fish. In order to empower the sector beset with constraints, interventions have to be applied holistically as well as specifically. Along with the increasing of the number of breeders and assistance schemes, participatory rural appraisals have to be planned for the sector. The need for the day is not only to enable the breeders for increased production but to link them to the production and supply chain supplying highest quality products. Strong linkages of the network can promote production efficiency, productivity growth, technological and managerial capabilities and market diversification. Activities of the State fisheries in Tamil Nadu, Tamil Nadu Fisheries Development Corporation (TNFDC), M.S. Swaminathan Foundation, Marine Products Export Development Authority (MPEDA) and Popular Non Governmental organisations such as Dhan and the State Agricultural Universities (TANUVAS TNAU, etc.) have to join hands to the development of ornamental fisheries in the state and in the country.

Acknowledgment

The funding provided by UNCTAD and MPEDA for the conduct of the Project on "*Developing Strategies to Network the ornamental fish Breeders in India for Enhancing Exports*" is greatly acknowledged. The Director, School of Industrial Fisheries is acknowleged for providing facilities for the conduct of the project.

References

Aaker, D.A., Kumar, V. and Day, G.S., 1997. *Marketing Research*, 6th edn. John Wiley and Sons Inc., New York, 776 pp.

Anon, 2009. *MPEDA Statistics*. Unpublished data.

Balasubramanian, M. and Eswaran, R., 2008. Marketing practices and problems of cotton cultivators in Virudhunagar district. *Indian Journal of Marketing*, 38(7): 27–32.

Dey, V.K., 1996. Ornamental fishes. *Handbook on Aquafarming*. Marine Products Export Development Authority, MPEDA House, Panampilly Avenue, Cochin, 75 pp.

Kumar, S.B.A., 2005. Import of live ornamental fishes and fishery products. In: *Ornamental Fish Export*. Cochin, Marine Products Export Development Authority and INFOFISH, Cochin, India, pp. 6–8.

Mahalakshmi, 2008. Cost and return in vanilla cultivation: A study with special reference to Coimbatore district. *Indian Journal of Marketing*, 38(3): 47–53.

Perumal, M., Mohan, R.J., Mohideen Raja, O.M. and Latasri, O.T.V., 2008. Marketing strategies of perishable commodities with special reference to flowers in Srirangam market, Thrichirappalli, Tamil Nadu. *Indian Journal of Marketing,* pp. 28–32.

Sekharan, M.N. and Ramachandran, 2006. Threats in exporting ornamental fishes from India to Singapore. *Seafood Export Journal.*

Sekharan, M.N. and Ramachandran, 2007. Bottlenecks encountered in exporting ornamental fishes from India. *Asian Fisheries Forum,* Cochin, India.

Sekharan, M.N. and Ramachandran, 2008. A probe into the constraints in exporting ornamental fishes from India. *Ornamentals Kerala–2006.*

Sekharan, M.N., 2008. Report Submitted to MPEDA–UNCTAD on Developing Strategies to Network Ornamental Fish Breeders in India for Enhancing Exports, pp. 199.

Sekharan, M.N., 2010. What hampers the ornamental fish production from India? *Souvenir, Ornamentals Kerala–2010.*

Slevaraj, A. and Vijaysanthi, K.R., 2009. Marketing problems faced by Mulberry growers: A study in Erode district of Tamil Nadu. *Indian Journal of Marketing,* 39(5).

Vasanthi, S., 2008. Challenges faced by cutflower growers in Tamil Nadu with special reference to Nilgiris district. *Indian Journal of Marketing,* 38(8): 15–21.

Index

A

Acanthocobitis moreh, 59
Acanthuridae, 181
Acclimatization, 208
Acid or alkaline water, 150
Agriculture Veterinary Authority (AVA), 26
Air diffusers, 125
Airlift pumps, 124
Amphiprion clarkii, 207
Amphiprion ocellaris, 193
Angelfishes, 187
Antennariidae, 182
Aplocheilus lineatus, 109
Aplocheilus panchax, 109
Apogonidae, 182
Aqua estate, 7
Aqua hub site, 32
Aquaponics, 127
Aquariculture, 3
Aquatechnology park (ATP), 32
Arowana, 138
Arulibarb, 16
Astaxanthin, 203

B

Badis badis, 105
Balistidae, 182
Barilius bakeri, 44
Barilius barna, 109
Barilius bendelisis, 109
Barilius shacra, 109
Bat fishes, 187
Biological oxygen demand (BOD), 130
Biopsy, 218
Biosecurity, 155
Blowers, 124
Botia almorhae, 103

Botia Dario, 104
Botia geto, 104
Bottom Manifold System, 124
Box fish, 187
Brachionus plicatilis, 211
Brachydanio rerio, 108
Butterfly fish, 15, 183

C

Caesionidae, 182
Carangidae, 182
Carbon filter, 170
Cardinal fish, 183
Carp lice, 152
Central Inland Fisheries Research Institute (CIFRI), 28
Central Institute of Freshwater Aquaculture (CIFA), 29
Chaetodontidae, 183
Chela fasciata, 45
Chlorella, 212
CITES certification, 33
Clown anemone fish, 193
Clown Fish, 171
Colisa chuna, 106
Colisa fasciata, 106
Colisa lalia, 106
Conditioning, 197
Contingency planning, 160
Conway medium, 212
Courtship, 225
Cow fish, 187
Ctenops nobilis, 106

D

Damsel fish, 15, 188
Danio aequipinnatus, 108
Danio dangila, 108
Danio devario, 107
Danio malabaricus, 108
Danio malabaricus (Jerdon), 46
Dermocystidium, 153
Detection, 159
Diodontidae, 183
Disease surveillance, 159
Domestic trade, 5
Dragnet, 180
Dropsy, 153

E

Echeneidae, 184
Eel, 186
Emergency preparedness, 160
Ephippidae, 187
Ergasilus, 152
Esomus danricus, 107
Etroplus maculatus, 215
Etroplus suratensis, 215
Exotic varieties, 5

F

Filament barb, 16
Filefish, 186
Fin rot, 151,153
Fish trade, 163
Fish trap, 179
Food and Agricultural Organization (FAO), 157

Frogfish, 182

Fusilier fish, 182

G

Gagata cenia, 104

Gagata itchkeea, 104

Garra mullya, 58

Germplasm piracy, 9

Gill parasite, 152

Glass fish, 16

Global Marine Aquarium Database (GMAD), 164

Globe fish, 190

Goat fish, 15, 186

Green certification, 31

Greenhouse facility, 117

Green water, 201

Grouper, 189

Grunt, 186

Grunter, 189

Gulf of Kutch, 15

Gulf of Mannar, 15

Guppy, 10

H

Haemulidae, 185

Hatching nest, 199

Health certification, 158

Hemirhamphus leucopterus, 105

Hemirhamphus limbutus, 105

Hemirhamphus xanthopterus, 105

Heterotidinae, 140

High density polyethylene (HDPE), 116

Hi-tech Public Aquarium, 7

Holocentridae, 185

Hook and line, 180

Hormonal manipulation, 216

Human Chorionic Gonadotrophin (HCG), 225

Hydroponics, 129, 130

I

IAMWARM, 7, 231

Ich disease, 151

Ichtyophonus, 153

Incubation, 210

Inspection, 158

International Council for the Explorations of the Sea (ICES), 157

International Maritime Organizations (IMO), 157

International Union for the Conservation of Nature, 157

Intramuscular implantation, 223

Isolation and observation, 159

K

Kerala Aquaventures International Ltd. (KAVIL), 32

Koi herpes virus, 154

Koolie barb, 16

L

Labeo calbasu, 110

Labridae, 185

Lavbuca dadiburjor, 108

Legislation, 156

Lepidocephalus thermalis, 65

Lernaea, 152

Leuteinizing Hormone Releasing Hormone – analogue (LHRH – a), 225

Lift traps, 179

Lighting facility, 125

Lining material, 116

Lionfish, 188

Low density polyethylene (LDPE), 115,116

Lutjanidae, 185

M

MAC certification, 168

Macropodus cupanus, 106

Marine Aquarium Council (MAC), 168

Marine Ornamental Fish Hatcheries, 7

Marine Products Export Development Authority (MPEDA), 44

Mastacembelus armatus, 107

Mastacembelus pancalus, 107

Melon barb, 16

Mesonemacheilus triangularis, 64

17-µ methyl testosterone (MT), 223

Mobile Aquaritech Lab, 7

Monacanthidae, 186

Monodactylidae, 186

Monodactylus argenteus, 215

Moony fish, 186

Moorish idol, 190

MS – 222, 222

Mullidae, 186

Multiport valves, 120

Mystus tengara, 104

N

NADP, 7

Namacheilus evezardi, 104

Nandus nandus, 105

Nangra punctata, 105

Nangra viridiscens, 105

Nanochloropsis, 212

National Bureau of Fish Genetic Resources Institute (NBFGRI), 42

National Fisheries Development Board (NFDB), 7, 233

Nemacheilus monilis, 60

Nemipteridae, 186

Neon tetra, 10

Nutrient film technique (NFT), 128

O

Odinium infection, 152

Official international des Epizootics (OIE), 31

Ophiocephalus punctatus, 107

Ornamental Fish Trade, 33

Osteobrama bakeri, 48

Osteochilus (Osteochilichthys) *nashii,* 56

Osteoglossidae, 140

Osteoglossinae, 140

Ostraciidae, 187

Ovaprim, 224

Ovatide, 224

Oxygen injection, 120

P

Packing technology, 9
Pair formation, 206
Parental care, 211
Parrot fish, 15, 188
Pellet implantation, 223
Pempheridae, 187
2-phenoxy ethanol, 222
Pipe fish, 189
Plistophora, 152
Polyethylene canula, 222
Pomacanthidae, 187
Pomacentridae, 188
Porcupine fish, 183
Premnas biaculeatus, 193
Prisolepis marginata, 69
Pristolepis fasciata, 68
Protein skimmer, 170
Puffer, 190
Puntius amphibius, 49
Puntius arulius, 99
Puntius arulius arulius, 50
Puntius chola, 99
Puntius conchonius, 99
Puntius denisonii, 99
Puntius denisonii (Miss Kerala), 51
Puntius filamentosus, 53, 99
Puntius gelius, 99
Puntius jerdoni, 55
Puntius melanampyx, 100
Puntius narayani, 100
Puntius phutunio, 100
Puntius sarana subnasutus, 55
Puntius terio, 100
Puntius ticto, 103

Q

Quarantine, 157, 158, 234

R

Rabbit fish, 189
Raceway Outlets and Filter Pipes, 119
Raceway tanks, 115
Rapid sand Filters, 120
Rasbora daniconius, 107
Recirculation, 221
Red line torpedo fish, 16
Remoras or suckerfish, 184
Reservoir pond, 117
Rhynchobolella aculeata, 107
Risk analysis, 158
Rohtee cotio, 109
Rosy barb, 16

S

Saprolegnia, 153
Scad, 182
Scaridae, 188
Scat, 188
Scatophagidae, 188
Scatophagus argus, 67, 215
Schistura denisoni denisoni, 62
Scleropages aureus, 141
Scleropages formosus, 141
Scleropages legendrei, 141
Scleropages leichardtii, 142
Scleropages macrocephalus, 142

Scorpaenidae, 188
Seahorse, 189
Serranidae, 189
Siganidae, 189
Skin / scuba diving, 180
Skin lesion, 153
Small scale industries (SSI), 243
Snappers, 185
Somileptes gongota, 104
Spinecheek anemone fish, 193
Spotted scat, 217
Squirrel fish, 185
Surgeon fish, 15, 181
Sweeper fish, 187
Sygnathidae, 189
Syngnathus kalyanensis, 110
Synthetic fibre spread, 116

T

Tagging, 219
Takifugus ocellatus, 215
Tangs, 181
Teflon sheet, 117
Temperature control, 125
Tetradon sp, 215
Tetraodon travancoricus, 67
Tetraodontidae, 190
Tetrodon cutcutia, 110
Tetrodon fluviatilis, 110
Therapon jarbua, 215
Theraponidae, 189
Threadfin breams, 186
Ticto barb, 16
Tiger fish, 189
TNFDC, 5
Toxotes chatareus, 215
Toxotes jaculatrix, 215
Trigger fish, 182

U

United Nations (UN), 157
United Nations Conference on Trade and Development (UNCTAD), 44, 236
UV lighting system, 170

V

Vegetable-hydroponics, 130
Velvet disease, 151

W

Western Ghats, 5
Western Ghats Development Programme (WGDP), 232
WGDP, 7
World trade organization (WTO), 157
Worms, 152
Wrasses, 15, 185

Y

Yellow cat fish, 16

Z

Zanclidae, 190
Zoning, 160
Zoonosis, 154

www.ingramcontent.com/pod-product-compliance
Lightning Source LLC
Chambersburg PA
CBHW070706190326
41458CB00004B/864
9789351241317